Small, Gritty, and Green

Urban and Industrial Environments

Series editor: Robert Gottlieb, Henry R. Luce Professor of Urban and Environmental Policy, Occidental College

For a complete list of books published in this series, please see the back of the book.

Small, Gritty, and Green

The Promise of America's Smaller Industrial Cities in a Low-Carbon World

Catherine Tumber

The MIT Press
Cambridge, Massachusetts
London, England

First MIT Press paperback edition, 2013

For information about special quantity discounts, please email special_sales@mitpress.mit.edu

This book was set in Sabon by Graphic Composition, Inc., Bogart, Georgia. Printed and bound in the United States of America.

Library of Congress Cataloging-in-Publication Data

Tumber, Catherine.
Small, gritty, and green : the promise of America's smaller industrial cities in a low-carbon world / Catherine Tumber.
p. cm. — (Urban and industrial environments)
Includes bibliographical references and index.
ISBN 978-0-262-01669-8 (hardcover : alk. paper)— 978-0-262-52531-2 (pb.)
1. Cities and towns—Environmental aspects—United States. 2. Fossil fuels—Environmental aspects—United States. 3. Land use, Urban—Environmental aspects—United States. 4. Renewable energy sources—Technological innovations—United States. 5. Technology—Economic aspects—United States. I. Title.
HT153.T86 2012
333.770983—dc22

2011011788

10 9 8 7 6 5 4 3 2

To my parents, William and Janet Stover Tumber

The city is the ideal state, the mean between the tyranny of the village and the tyranny of the metropolis.

—Léon Krier

After you leave New York, everything else is Bridgeport.

—George M. Cohan

Contents

Acknowledgments

The germ of this book lay in my longstanding alarm over what was happening to the communities—the smaller industrial cities—in which I'd spent most of my life. I knew that one day I would write about it. I have lived long enough to witness not just their deterioration but also their growing invisibility as cities—as further evidenced by several tepid responses I received to my pitches for an article on the subject. So I owe my first debt of gratitude to editors Deb Chasman and Josh Cohen, who had the imagination and intellectual fortitude to publish the essay on which this book is based in the *Boston Review*.

Once the essay saw the light of day, several people urged me to expand my thoughts into a book. Martha Bayles and Susan Linn really pressed the case, and their enthusiastic encouragement pushed me past the edge of indecision. Thanks a lot, you guys: I just lost a year and a half of my life. Clay Morgan, my editor at MIT Press, had the same idea, and before long, I was knee deep in this project. I thank him for ushering me through the process with aplomb and for arranging a grant from the Furthermore Foundation to support my efforts. I am, of course, extremely grateful to Furthermore, whose early and generous support made it unnecessary for me spend precious time on fundraising from other sources. Urban and Industrial Environments series editor Robert Gottlieb's encouragement was indispensable to the writing process, as were his editorial and substantive comments, along with those of several peer reviewers. They saved me from several errors of fact and judgment, and I learned a great deal from their reports. Indeed, this is probably as good a place as any to say that I alone am responsible for whatever weaknesses trouble these pages.

Early on I was blessed by the kindness of strangers, without whose generosity I cannot imagine having gotten this project off the ground, much less completing it. Dayna Cunningham and Amy Melissa Stitely, of the MIT Department of Urban Studies and Planning's Community Innovators Lab, arranged to provide me with an ongoing research affiliate appointment in fall 2009. As a result, I have had access to the tremendous resources of the MIT Libraries system, as well as to students, staff, and outside speakers from whom I have learned much along the way. James J. Connolly invited me to present a paper at the Center for Middletown Studies in spring 2009, even before I had a book contract, which introduced me to the rarified world of small-city scholars. Those who attended are also very much in my debt, especially Amy Scott, Allen Dieterich-Ward, and Eric Sandweiss. Since then, Jim has been unstinting in his collegial support while working feverishly on a film documentary and two books (including *After the Factory: Reinventing America's Industrial Small Cities*, a volume of our 2009 conference papers). He even found time to read a first draft of my manuscript. And what would I have done without Joe Schilling, of Virginia Tech's Metropolitan Institute and the National Vacant Properties Campaign? I shudder to think. From the moment he read my original essay, which aligned with much of his own thinking, he morphed into a one-man contact machine—sending it around to his seemingly limitless bounty of colleagues working on all aspects of older-industrial-city policy and keeping me abreast of developments in the field. And when it came time for me to delve into reporting, with little time to waste, he e-introduced me to many of my first contacts in the field. Thank you, Joe, for all that, and also for reading my manuscript with care.

I wish I had the space to thank every one of the many people who shared their time, ideas, insights, and cities with me as I did the research and reporting for this book. Only a small fraction of those I spoke with are included in these pages—but I am deeply grateful to all of them. I wish I could have included more stories and covered more aspects of the cities I do discuss. Everything I heard and saw was profoundly enlightening and often moving, and forms critical background to the ideas I present here. And then there were the people who not only talked with me, but also showed me around or introduced me to many others: Landon Bartley, Paula Davis, Bill D'Avignon, Steve Filmanowicz, Ben Forman, Jeramey Jannene, Jonah Katz, Jim Lively, Kurt Mulhauser, John Palmieri, Dave Reed, Jajean Rose-Burney, Nancy Rosin, Terry Schwarz, Christopher Setti, Robert G. Shibley, Amy Simon, Greg Spiess, and Benjamin Walsh. Thank you for sharing your worlds with me.

Linda Luz-Alterman occupies a special place in my heart as I reflect on how this book came about. To the late Christopher Lasch, I am indebted in ways I can hardly describe. I don't know how I would have made it through this crazy world thus far without his rich social criticism—even that with which I disagreed—and exemplary intellectual courage.

One of my friends, who shall remain nameless, has dubbed my book "Small Cities: Why You Should Give a Sh** about Them." I am so lucky to have friends like him, who have kept me in good humor and serious conversation, as needed—even when they did not quite understand the depth of my affection for smaller industrial cities. These friendships mean the world to me, and so it gives me great pleasure to acknowledge Angela Amato, Cheryl Bennett, David Chappell, Jon Garelick and Clea Simon, Judy Greene, Rochelle Gurstein and Jack Barth, Dan Kennedy, Sean Kenniff, Betsy and Ray Lasch-Quinn, Chris Lehmann, Colin Morris, Cathy Polovina, George Scialabba, John and Anna Summers, Jim Tumber, and JoAnna Wool. You know what you've done, and I thank you for doing it.

My parents, William and Janet Tumber, are extraordinary people—everyday heroes, who emerged from compromised circumstances to create whole lives together over time, a long time. Astonishingly, they keep growing into themselves by dint of their curiosity, courage, and plain old practicality, and so they go on loving the world ever more. What greater foundation can a child ask for? They have been behind me all the way on this book venture, as have my siblings—Julie, Jim, Bill, and Deb, along with their spouses—and my brood of seven nieces and nephews, who have cheered me along without having the faintest idea what I was writing about.

Finally, I thank My Favorite Marjie—Marjorie Siegel—who read successive waves of my manuscript with care and kept me grounded from afar as I toiled away at my desk. Here's to spending a little more time together.

Introduction: Beloved Communities, Benighted Times

Growing up in a small upstate New York farming village, I regularly took bus trips with my mom and little sister into "the city": Syracuse. Like most other non-urban white middle-class families in the early 1960s, we had one car, which my dad drove to work. So we would buy our tickets at the village pharmacy, board the bus, and barrel though miles of farmsteads and sparsely developed land until we reached the short stretch of highway into town. A half-hour after departing we would disembark in Syracuse's vibrant downtown, all glittering lights and vertical planes, filled with haberdashers and department stores, jewelry and candy shops, movie palaces, "ethnic" restaurants, and city people who were interestingly not like us. Although it might be hard to imagine for those whose memories don't stretch back that far, downtown Syracuse—like downtowns in most other postwar smaller industrial cities—had much the feel of today's midtown Manhattan. *Don't laugh.*

That world is now gone, gutted by a ghastly downtown interstate, the dispersion of retail to suburban malls, and white middle-class flight. The deadly forces of deindustrialization, outsourcing, and globalization issued the coup de grace and gradually emptied the city of jobs and population. The demolition of once stately commercial and public buildings, lovingly erected by local patrons over the course of a century, soon followed. My rural hometown, for its part, was long ago folded into the hideous aesthetic logic of suburban sprawl—what James Howard Kunstler has dubbed "the geography of nowhere."[1]

Cast your eye across nearly every smaller industrial city in the American Northeast and Midwest today, and you find squalor worthy of a Third World country. Appalling levels of poverty. Lousy school systems filled with historically neglected minorities and overburdened by non-English-speaking immigrants priced out of larger cities. Aging populations unable to retain their young. Tax bases in free fall. Deteriorating infrastructure, broken streets, abandoned buildings, and boarded-up houses ransacked by metal thieves. Some of these places—most notoriously Youngstown, Ohio, and Flint, Michigan—have lost as much as half their populations. Here, as in Detroit and Cleveland, once teeming neighborhoods have been replaced by acres of rolling prairie, their settled pasts now only intimated by phantom tree patterns and ghostly driveway openings.

By "smaller," I mean small-to-midsize cities that at their peak in, generally, 1950, had populations of roughly 50,000 to 500,000 souls, and whose numbers today have dropped (though not universally) by at least 20 percent.[2] And I do mean "cities," not small towns. Invisibility has been an overarching indignity these once-thriving urban centers have suffered in a postindustrial culture that, rhetorically and imaginatively, divides the world between the metropolis (Wall Street) and the small town (Main Street). That invisibility has provided cover for the very real urban troubles besetting these smaller metros.

For all that, this is a hopeful book. I argue that smaller industrial cities, long ignored and even maligned by urban theorists, could have a bright future if they prepare now for a low-carbon world—a world in which they could play a central role. My claim is based on two related convictions. First, we must end our dependence on fossil fuels. Second, our long-term environmental well-being depends on dramatically altering modern land use patterns—in agriculture, in transportation, in housing, and in workplaces and retail establishments—by concentrating population in cities and inhibiting sprawl. Most scientists agree that global warming poses a serious threat; even civic and business leaders who don't believe climate change is man-made are increasingly alive to the geopolitical necessity of limiting our dependence on fossil fuel.[3] I assume here that the high-carbon end game is upon us. The second notion, however, is badly in need of refinement in ways that take into account the promise of smaller industrial cities *as a class*, with distinct attributes that large cities lack.

Small-scale urbanism could be a virtue and a strength in the emerging low-carbon economy for three reasons. Even in their diminished state, smaller cities have population density and the capacity for much more. They also have land assets most large cities lack: within municipal borders, in more sparsely settled suburbs, and in more closely proximate and abundant open space. In the Northeast and particularly in the Midwest, that land is among the richest in the world. It's a point that is too often overlooked: in a sustainable future, land will be needed for relocalizing agriculture, siting windmills, solar farms, anaerobic digesters, and other low-carbon energy harvesters, and growing raw vegetable material for biomass production.

Finally, smaller cities in these regions have manufacturing infrastructure and workforce skills that can be retooled for the production of renewable technologies, such as clean-fuel automotives, trains, and windmill, solar, and hydropower components—for which there will be real market demand in a low-carbon economy. We are, after all, on the brink of a third industrial revolution, and these cities are highly suited to play a central part in it.

Greening the metropolis with community gardens, farmers' markets, green roofs, and ecologically balanced waste management is a fine thing. So are weatherizing current structures and erecting new energy self-sufficient green buildings. Smaller cities should take these measures too—all cities should, if we are to reduce greenhouse gas emissions to a responsible level. After all, cities worldwide consume 75 percent of the world's energy and produce 80 percent of its greenhouse gas emissions.[4] But the *productive* green economy is going to require land resources near densely settled areas, which, in combination, global cities like New York simply don't have. The United States is in a position to lead and prosper from this *new* new economy if it acts wisely and swiftly. Smaller cities in the Northeast and Midwest could—indeed, must—play a central, if decentralized, role in this transformation and, in the process, reframe the very ways we think and talk about urbanism.

Placing value on smaller-scale urbanism and arguing for its critical position in a low-carbon future does not have to come at the expense of large cities. If anything, across the ages and continents, both types of urban settlements have needed each other, and there is no reason to think that that symbiosis will or should disappear.

Scale does matter a great deal, however. There is much talk among urban-and-economic-policy experts about further concentrating population in megaregions stretching between large cities such as Boston and Washington, D.C., Los Angeles and San Francisco, or Charlotte and Atlanta, tied together by sophisticated high-speed rail. As compelling as it is, this vision of transportation planning does not factor in the spatial and resource requirements of the productive green economy—something for which the web of smaller cities in the Northeast and Midwest is historically and geographically well suited. Besides, not everyone who thrives on the sociability and imminent mood of urban culture wants to live in New York or Chicago, Los Angeles or San Francisco, even if they could afford to do so. Enormous metropolitan areas can be ungainly and difficult to negotiate on a number of levels. No matter how sophisticated their public transportation systems, they require countless hours in transit because of their sheer magnitude. Their rising urban land values require a pitch of density even sworn city lovers can find excessive. And it can be difficult for ambitious young adults to gain a political or economic foothold in big cities, where the power centers are mighty and well established in nearly impenetrable social networks generations in the making.

Likewise, many, if not most, suburban dwellers feel as though they have been exiled, forced into a state of auto-dependent gloom to live near the exurban office parks where they are employed and to put their children in relatively functional public schools.[5] As one reluctant suburbanite commented in a Gristmill blog discussion: "I think everyone is missing the mark—most suburb and exurb dwellers are often people who would prefer to live in small to mid-size cities, and some even prefer small towns, but in this economy we have to go where the jobs are. . . . If we bring jobs and opportunity back to small cities who need the growth, we can relieve the pressure on major metropolitan areas that have sprawled too far."[6]

In the United States, smaller industrial cities are generally regarded with condescension, if they are seen at all down there in "flyover country." It's fair to say that what urban historian James J. Connolly calls "metropolitan bias," more than forty years in the making, has taken root in American life. By the 1970s, when the last book-length studies of small cities were published, they vanished from discrete consideration, relegated to "best-of" lists or the longstanding community studies tradition. To date, no one has created a comprehensive bibliography of small city studies as a class, which itself underscores the point.[7] Over the past decade, a small scholarly and policy literature concerned with the history and current plight of smaller cities has quietly taken shape. Nearly all of them remark that these studies are nascent and conclude with calls to action.[8]

Students of urbanism have long assumed that "the beginning of what is distinctively modern in our civilization is best signalized by the growth of great cities," as Louis Wirth put it in a still influential 1938 essay, "Urbanism as a Way of Life."[9] In other words, early on, the study of cities became a means of understanding the process of modernization, and the larger and more central the city in a hierarchy of cities, the more exemplary it was for the study of modern politics, psychology, economics, and sociology. The same held true for the revisionist new urban history of the 1960s and 1970s, which was primarily concerned with kinship structure, class formation, and social mobility. Even the cultural postmodern turn in history, bent on decentering received hierarchies and empowering the margins, focused on finding alternatives to nationalism in the multicultural cosmopolitan communities of the metropolis. The idea that smaller cities themselves had been marginalized or that any of these processes might work differently in cities of smaller scale never seemed to arise during these decades of scholarship. Although that is now beginning to change ever so slightly, scholars, Connolly observes, generally "treat the metropolis as the quintessential urban form."[10]

Metropolitan bias, left largely unquestioned, pervades the work of today's most influential American urban theorists. All cut their intellectual teeth on the legacy of Jane Jacobs, whose work, beginning with *The Death and Life of Great American Cities* (1961), inspired a reappreciation of big cities and their neighborhoods after years of intensive suburban development. She also bequeathed to later urbanists her disparagement of "little cities" and "dull" factory towns. Today's urban-economic tastemaker, Richard Florida, urges cities to build walkable communities with amenities that appeal to what he calls the "creative class"—the innovative knowledge workers and professionals who make the global economy hum and are drawn to the open-minded restlessness of artists and "bohemians," the young and the "gay"—rather like Jacobs's West Village. Florida's one-model-fits-all-sizes approach applies to smaller cities too, but mainly to university towns teeming with creative class mojo. His global cosmopolitanism has grown only harsher with his recent work, *The Great Reset* (2010), on the long-term economic shakeout and "spatial fix" that got under way with the 2008 recession. As "talent" and economic growth concentrate further in megaregions anchored by global cities, Florida argues, smaller industrial cities—unable to harness "The Velocity of You," as he titles one of his chapters—are best left to fade away, unfortunate casualties of the irresistible forces of globalization. Harvard economist Edward Glaeser, who writes the popular *New York Times* Economix blog, is also enthusiastic about megaregional growth, arguing that globalization has led to "bigger cities" because "smart people" need to "hang out" together. Joel Kotkin, an aggressive defender of suburban form, models his work on large metro suburban areas without considering that sprawl might take a different form in smaller cities. More recently, projecting the consequences of population growth in *The Next Hundred Million: America in 2050*, he anticipates growth in the heartland and its small cities, but only because they have space—not because of any inherent value in smaller urban scale. Joel Garreau, writing about exurban "edge cities," and Alan Erhenhalt, who has been studying the migration of the working poor to first-ring suburbs, inverting the "doughnut effect," confine their work to the study of large metro areas. Meanwhile, the smart growth and new urbanist movements, both attentive to the environmental consequences of sprawl, generally argue for models of sustainability that work on all scales. But few consider the ways in which small urban scale—and I don't mean small towns—might be an advantage and a strength in a low-carbon economy.[11]

Some smaller cities have attracted newcomers in recent years. Empty nesters and retirees, as well as a sprinkling of young adults, are drawn to university towns and cities such as Mankato, Minnesota, and Fond du Lac, Wisconsin, by the cultural and recreational amenities they provide. But even here, popular culture cannot quite bring itself to call them what they are. On one of my research trips for this book, I picked up an in-flight travel magazine with a cover story on Asheville, North Carolina. "This small mountain burg" (of some 78,000 people, it turns out) has "that big city/small town synergy," it observed. One resident reported, "Asheville reminds me of being in a bigger city . . . but not. If you moved [New York's] East Village and put it in the mountains—that's Asheville."[12] And that's it: even small cities that are successful and have attracted favorable publicity barely come into focus as cities. Those in more dismal shape, struggling to make ends meet in the Rust Belt, do not get much of a hearing at all.

* * * * * * *

Beginning in fall 2008, I made several trips to smaller industrial cities throughout the Northeast and Midwest to learn how—and whether—local civic leaders and activists were putting their metros on a sustainable footing. It was, in a sense, a homecoming for me writ large. Although family matters brought me to Boston, where I now live, I spent most of my life in the so-called Rust Belt—mainly in the smaller industrial cities of Syracuse and Rochester, but also in Albany and Detroit. I know what it's like to watch downtown die and to see city leaders try to revive it with stadiums and riverfront development while predominantly poor neighborhoods and their schools languish. I know what it's like to see malls steal retail from the city, leaving the old shopping districts to wither. I know what it's like to be asked, over and over, "Why do you want to live *there*?" I know what it's like to have to drive out to the suburbs, with steam coming out of my ears, to find a grocery store. I know what it's like to have to drive everywhere because of inadequate mass transit.

But I also know the joy of getting out into the countryside just fifteen or so miles from the city center, past the big-box stores, and encountering an older, if more conservative, agrarian culture grounded in its own sense of rural beauty and pride in work—one that resents the city but, here and there, longs to be better integrated with it. I've enjoyed strolling through city neighborhoods in various states of repair or dishevelment—my own and others'—looking for layered clues to their histories. My bland small-city palate has traveled a long journey from Spam and diner food—maybe a little Italian and Chinese—to the tastes of Southeast Asia, Latin America, Africa, and Eastern Europe as I developed a growing appreciation of how these newer immigrants' ways added greater dimension and sparkle to the local culture. And I've known many people throughout the years who have fought hard for their small cities. I've engaged in a few of these battles myself.

This book is idiosyncratic, covering ideas and styles of inquiry usually kept separate, all in the service of a view at odds with conventional wisdom: a hopeful vision for America's small-to-midsize industrial cities. It reflects not only my love of these places, but also my background in the fields of history and journalism and my formative engagement with the environmentalist movement, with its broad insight that the human condition imposes moral and natural limits to growth. During the 1980s, I watched in horror as the Reagan Revolution pounded away at the fundamentally conservative impulses of the environmentalist movement while radically deregulating the market economy. It struck me as the height of hubris, and I saw little in the liberal left, which was preoccupied by identity politics and poststructuralist musings, to counter it. In graduate school I studied the late-nineteenth-century Gilded Age, a period much like our own, when the great trusts called the shots with minimal accountability and boundless ambition. My longstanding frustration with the disintegration of smaller cities is of a piece with this broader democratic skepticism of great concentrations of wealth and power and with a spiritual longing to restore a sense of appropriate scale to our ways of inhabiting and making sense of the world.

In all, I traveled to twenty-five cities, including the larger ones of Detroit, Cleveland, Buffalo, and Milwaukee. Beyond a few stops in New England and upstate New York, time constraints forced me to confine my travels to Michigan, Ohio, Illinois, Indiana, and Wisconsin—much as I would have liked to press farther west and into Pennsylvania. Some of the cities I visited (Grand Rapids, Akron) made the transition to a postindustrial economy faster and more cleverly than others; since they never really hit bottom, their urban fabric and neighborhoods remain relatively intact. Others (Syracuse, Rockford, Canton) are working frantically to make up for lost time. Still others (Flint, Youngstown) are commonly referred to as basket cases. Like the central cities of which they were once satellites,

Detroit and Cleveland, they have lost so much population and are in such horrendous fiscal straits that they can barely keep pace with the demand for uninhabitable-housing teardowns. To describe these neighborhoods, especially in contrast to their condition thirty years ago, is like trying to convey the majesty of the Grand Canyon: the sheer scale of the devastation must be seen personally to compass. As a result, they must completely reinvent themselves, and even with minimal resources, they are already hatching some of the most compelling experiments in sustainability.

There was little design to the way I selected which cities to visit, aside from considering their size and manufacturing histories. Partly it was a matter of whom trustworthy contacts could introduce me to, and sometimes, when I had to go it alone, it was a matter of who was willing to talk with me—a perfect stranger writing a book. Once in town, I followed my leads. My jerry-rigged method put me in touch with a spectrum of approaches from a variety of perspectives. Some of the folks I talked with were city and regional planners principally concerned with land use planning and zoning. Others were responsible for local economic development or for charting out state policy. A few were city council members and mayors or city managers. Still others were academic experts in sustainability or were affiliated with universities' efforts to work with their host cities to cultivate community and economic development. And then there were the untold citizen activists—paid community organizers, community activists, newspaper editors and writers, bloggers, and the occasional unclassifiable iconoclast. I also spoke with several developers and entrepreneurs and a host of farmers and market organizers. As the book unfolded, I learned yet more from secondary accounts and through telephone interviews. Some are braving tremendous odds to fight sprawl and alter our transportation system and, in the process, trying to repurpose older industrial infrastructure, or "brownfields," to suit today's needs. Others are exploring alternative forms of agriculture or energy production, or ways of retaining local community wealth. Still others are fighting a painful uphill battle for social equity in impoverished, predominantly black and Latino inner cities surrounded by affluent white suburbs, particularly as it has played out in the public schools. Sometimes they are at cross-purposes with one another, often they are only dimly aware of one another's efforts, and their politics are all over the map. Their stories are arresting, amusing, sometimes strange, and often inspiring. Fascinating and unpredictable too are the tales of those who more or less defend the status quo or mask it with "greenwashing"—marketing environmentally dubious practices with eco-sustainable messaging.

In the chapters that follow, I weave their stories into an exploration of ideas, both old and new, that value smaller urban form. This book is intended to be suggestive and broad rather than penetrating and thorough, alive to the strengths of smaller urban scale across a range of low-carbon practices and economic development strategies. It is hardly a comprehensive, definitive study of the sort that specialists undertake after years of fieldwork and bibliographical research. I would count it a great honor if it inspired a few theses, articles, and case studies—not to mention more books—that put to the acid test ideas I only touch on here across the fields of urban planning and theory, American cultural and political history, agriculture and renewable energy technology, public education and retail development. I would count it an even greater honor if it broadened the knowledge of citizens working diligently to revitalize their smaller cities and inspired them to press for state and federal policies that recognize how crucial their metros are—again, as a class—to a sustainable future.

* * * * * * *

Small-to-midsize older industrial cities are up against formidable odds. It doesn't help that their place in American cultural geography has always been somewhat ephemeral and confusing. Coming into existence with the birth pangs of the American Century, smaller industrial cities were moving targets and hard to define, beholden to post–Civil War railroad networks in search of market expansion that either induced growth or spelled doom in older towns. As a result, until the 1910s their sizes were unstable, and they were difficult to conceptualize: Were they overgrown towns, or were they underdeveloped cities?[13]

Even at the peak of their industrial might, most were "peripheral" or "satellite" cities, dependent on the automobile, steel, and agricultural commodities powerhouses of Detroit, Cleveland and Pittsburgh, and Chicago. Many were one- or two-company towns, poised precariously to rise or fall with decisions made hundreds, even thousands, of miles away. The well-being of agriculture too—a significant economic driver in most smaller industrial cities—was dependent on credit and tariff policies over which farmers had little control. From their modern inception, smaller industrial cities were the product of market systems constructed by eastern bankers and investors whose legacy of control their citizens resent to this day.[14]

Obviously the decline of industry played a huge structural role in these cities' harsh fortunes. Some, such as Akron, Toledo, and Lima, in Ohio,

Muncie, Anderson, and Kokomo in Indiana, did well for a while, particularly in the auto industry as Detroit began in the late 1950s to outsource its manufacturing beyond the unionized plants of the Motor City to Detroit's suburbs and to smaller cities and rural areas in the Midwest and South. But by the 1970s, when steel and the Big Three faced stiff competition overseas and trade protections were further weakened, manufacturing in everything from household electronics and the toy industry to the garment and footwear trades began moving offshore, where labor costs were low and environmental protections (newly enacted in the United States) were minimal to nonexistent. Meanwhile, Japanese carmakers set up most of their American-market shops in the low-wage, nonunionized South. With the passage of the General Agreement on Tariffs and Trade and the North American Free Trade Agreement of the early 1990s, what began as a trickle became a wholesale exodus. Now American workers must compete not only with low-paid workers in distant lands, but also with immigrant laborers accustomed to a lower standard of living. Because of their peripheral state in the market system, smaller cities were hit particularly hard by these developments.

Deindustrialization does not tell the whole story, however. Before jobs began fleeing elsewhere in earnest, the character of the small industrial city as a distinct urban form had been damaged severely. By the late 1950s, as the national highway system began replacing rail transportation, highway planners left their scale out of the balance. The same downtown highway system model conceived by Robert Moses for New York and other large metros sliced through places like Syracuse, Akron, and Flint to much greater effect. Their smaller size left them less equipped to absorb the horrific consequences of downtown highway construction and the suburban retail it facilitated. Postwar urban renewal policies (truly an Orwellian term) had similarly disproportionate consequences: whereas in large cities these developments wiped out entire neighborhoods, which was disastrous and unfair in itself, smaller cities were altered in their entirety.

Yet before the 1960s, considerable intellectual energy went into creating smaller industrial cities and understanding their place in American culture. They arose simultaneously with the City Beautiful movement, the country's first, and distinctly American, urban planning movement, which flourished in the late 1890s and 1900s. Indebted to the public-minded naturalism of Frederick Law Olmsted Sr.'s urban park systems and the grandeur of Daniel Burnham's "White City" at the 1893 Chicago World's Fair, it is most commonly associated today with the National Mall in Washington, D.C. Yet the movement took root primarily in the small-to-midsize cities cropping up across the industrializing Midwest, as well as in new western cities. The City Beautiful came into flower as progressive municipal reformers grappled with the anarchy of explosive urban population growth, yet their aims were too modest and incremental to effect deep structural change in large, more established cities. City Beautiful advocates—mainly local elites joined in voluntary municipal art and civic improvement associations that served as informal planning boards—concerned themselves with the orderly grouping and placement of public buildings, railway stations, and parks, nurturing an exemplary vision of the urban public realm. Influenced by neoclassicism and the arts-and-crafts movement ideal that "what is most adapted to its purpose is most beautiful," they were particularly attentive to appropriate fit and scale. Some of their handiwork remains in smaller cities across the land, since the market for new downtown development, which usually results in the demolition of older buildings, did not take shape as it did in large cities over the past few decades. Preservationists have also worked hard to save them from the wrecking ball.[15]

By the 1910s, smaller cities had come into focus as an urban type, championed by landscape architect John Nolen, among the most influential early professional urban planners. A relentless advocate of smaller urban form deeply influenced by both the City Beautiful and the progressive civic awakening to the practical needs of the common welfare, Nolen argued that comprehensive planning could have "only narrowly limited influence in larger places . . . ameliorating merely the most acute forms of congestion, correcting but the gravest mistakes of the past." By contrast, he argued in his first book, *Replanning Small Cities* (1912), smaller cities, which at the time were growing faster than large ones, were in an optimal position to replan for the future. In addition to reforming financing arrangements for adequate schools, sanitation, and water services, Nolen exclaimed, "There is scarcely anything in the smaller places that may not be changed":

> In small cities, for example, railroad approaches may be set right; grade crossings eliminated; water-fronts redeemed for commerce or recreation, or both; open spaces acquired even in built-up sections. A satisfactory street plan can be carried out, and adequate highways established; public buildings can be grouped in at least an orderly way; and a park system, made up of well-distributed and well-balanced public grounds, can be outlined for gradual and systematic development. All of these civic elements, indispensable sooner or later to a progressive community, may be had in the small city with relative ease and at slight cost.[16]

Nolen's plans also sought to "establish the individuality of a city,—to catch its peculiar spirit, to preserve its distinctive flavor, to accent its particular situation," in contrast with the "cruel monotony" and "uniformity" of design brought about haphazardly by American commercial builders. "We should," in short, "have a local concept."[17] To these ends, Nolen, who undertook almost 400 commissions before his death in 1937, concentrated on planning smaller northern cities, from Reading, Pennsylvania, and Montclair, New Jersey, to Madison, Wisconsin, and Bridgeport, Connecticut, before turning his attention to projects in the South and West.[18]

By 1933, an influential survey of "the metropolitan community" recognized that a "hierarchy" of cities now existed in which the "problems of the large city" were distinct—and assigned a special section in the book. With the rise of technocratic modernism in metropolitan building design, the City Beautiful—both its aesthetic and its notion that cities are plannable—was scoffed at remorselessly as politically ineffectual and naive, the handiwork of Booster Club nitwits and their feminine counterparts. A 1934 review of regionalist Russell Van Nest Black's *Planning for the Small American City* captures the mood of derision: "Much of the literature on planning has the ring of fanciful idealism. The nontechnical public associates the term with city beautification and the lay members of planning commissions are scarcely more intelligent." In spite of such ridicule, consideration of "fitness and appropriateness" in urban form and architecture didn't go away. In part it took the shape, up through the 1960s, of lively debates among planners and urban theorists about the optimal size and arrangement of urban settlements, with regionalists Lewis Mumford, Benton McKaye, Clarence Stein, and Henry Wright paying the most careful attention to the value of smaller cities, as we shall see.[19]

As conveyed by Black's reviewer, however, no sooner had smaller cities made an appearance in the urban firmament than they were met with the cosmopolitan sneer mastered by H. L. Mencken, who viewed them as the natural habitat of *Boobus Americanus*: credulous, ignorant, crude, self-righteous, and "conformist." In the pages of the *Smart Set* and the *American Mercury*, which he edited in the 1910s and 1920s, Mencken's bemused contempt for democratic pieties descended on the small industrial city (and southern fundamentalist) with special force, setting the terms for its appraisal. From the Olympian heights of New York's publishing world, the new cosmopolitans not only made small cities a stand-in for the "imbecility" of middle-class business culture but also conflated these much larger settlements with the small town, setting up the Wall Street–Main Street dyad that has endured through the years.

Literary Midwesterners knew better and conceived the relationship more in terms of a triad, one that included "the Boulevard." In Iowan

Floyd Dell's semiautobiographical *Moon-Calf* (1920), the main character grows up moving from a small town and then to two small cities; in Fort Royal (Davenport) he finds a sense of self and community among a group of local writers and free spirits, as well as with his fellow laborers, before lighting out for Chicago in his thirties. Sherwood Anderson, best known for the village novel *Winesburg, Ohio* (1919), next portrayed the manufacturing city of Bidewell (population 100,000) in *Poor White* (1920), a conflicted account of the small city's veiled promise. It was the forced transience and deskilling of working people imposed by industry that most concerned these writers, not the city's oppressively small size or insularity.[20]

From a distance, it was all too easy to conflate the dramatically opposed demographic shifts taking place in the heartland. As small industrial cities were growing and the countryside was depopulating, a significant literature on the exodus from small towns had come into being. With varying degrees of respect, longing, guilt, and claustrophobia, writers such as Edgar Lee Masters, Sherwood Anderson, Booth Tarkington, and Willa Cather explored the small-town agrarian world that was being lost to the industrial dynamo. In 1921, critic Carl Van Doren christened this body of literature "The Revolt from the Village" and heralded Sinclair Lewis's *Main Street* (1920) its finest triumph. But Lewis took up the smaller city, too, and made it the focus of his next project. He conceived *Babbitt* (1922) as a portrait of one of those medium-size cities of "200,000 to 500,000," or "transitional metropolises" that, as he told his publisher, had never been "done" in American fiction. Lewis made good on his intent: "Cities of the type of Zenith," he wrote of the novel's setting, were "commercial cities of a few hundred thousand inhabitants, most of which—though not all—lay inland, against a background of cornfields and mines and of small towns which depended upon them for mortgage-loans, table manners, art, social philosophy and millinery." Lewis wanted to render George F. Babbitt sympathetically and, though his gift for mockery got the better of him, to a large extent he succeeded in capturing his main character's longings and self-doubt, his loneliness and thwarted aspirations, as he faced the existential futility of boosting and braying about real estate values. The novel also portrays a range of urban characters—poets, feminists, industrial socialists, bohemians—who would be out of place in a small town setting. Yet *Babbitt* is remembered mainly as a satirical portrait of the Menckenesque "boob" of American middle-class commercial culture. And while many contemporaries understood the difference, "Zenith" has been folded imaginatively into the small-town life

Lewis had depicted two years before in *Main Street*—a settlement of 7,000.[21]

Perhaps more than any other widely read book, Robert and Helen Lynd's sociological case studies of Muncie, Indiana—*Middletown*, published in 1929, followed by *Middletown in Transition* in 1937—also brought smaller cities into view. Again, though, they selected the small, midwestern, predominantly white settlement as representative of middle-class culture. In the first and more popular of the studies, theirs was at least a more affectionate and complex portrait of small-city life than that advanced by the cosmopolitan enemies of Babbittry. Concerned primarily with the effects of industrialization and mass culture on the city's business and family life, civic and religious organizations, and class arrangements, the Lynds harbored some respect for the denizens of Middletown. Much as they maintained a voice of studied sociological neutrality, they held out some hope for the capacity of its mainstream Protestant culture to resist the enforced passivity and standardization imposed by the mechanization of work and the cult of consumption. If Middletowners were "bewildered" by the speed with which the drive for uniformity from without had swept through their world since the 1880s, the Lynds implied, their dormant habits of "self-appraisal and self-criticism" could lead them eventually to "a reexamination of the institutions" to which they had conformed—that is, to both the insularity of the Rotary Club and the blandishments of radio, popular magazines, and movies. By the second volume, they were not so sanguine, but even here they painted a more complicated picture of life in a small city by laying bare the power of local economic elites—the Ball family—and exploring the lure of home-grown fascism among some of its working people and small businessmen.[22]

Sociologist C. Wright Mills also studied small cities, which he discussed at some length in two popular classics of social criticism: *White Collar* (1951) and *The Power Elite* (1956). With these works, the brilliant iconoclast of deeply conservative instincts analyzed what he saw as a new, more despotic society brought about by the expansion of corporate and military power during World War II. Mills's influence—his most lasting—can be seen in the New Left's critique of both the Communist Party and what they called the Liberal Establishment, with its penchant for perpetual war and "totalizing" bureaucracies. Lost in this legacy, however, is Mills's shrewd analysis of small-to-medium-size industrial cities of between 25,000 and 175,000 (which undoubtedly held little interest among civil rights activists of the New Left, who, with good reason, distrusted local power structures).

Much of Mills's discussion of small cities is couched in terms of "status anxiety," as the traditional elites of local society struggled to make sense of their irrelevance in the shadow of corporate domination. His deeper subject, however, was the shifting balance of power toward big cities. As the "national corporation has come into many of these smaller cities," Mills wrote in *The Power Elite*, "there have come the executives from the big city, who tend to dwarf and ignore local society." Local women "active in social and civic matters"—heirs of the earlier beautifiers and improvement leagues so belittled in Mencken's *Smart Set*—also lost power to the indifference of corporate-identified executives' wives. Meanwhile small cities lost much of their political autonomy while undergoing "gradual incorporation into a national system of power and status," as Mills put it. "Muncie, Indiana, is now much closer to Indianapolis and Chicago than it was fifty years ago." Power was no longer distributed among "decentralized little hierarchies." It now lay with transient salaried workers with little civic loyalty to place, who were more likely to live in the growing suburbs.[23]

The postwar period registered some of that uneasiness as urban experts, influenced by central place theory, debated the role of small cities in "systems of cities," mainly in terms of relative economies of scale and economic diversity.[24] Overall, though, suburbanization and its effect on "the city"—now meaning big cities—was rapidly changing the subject. By the late 1960s, older industrial cities of all sizes faced political and fiscal crisis, the bitter fruit of postwar suburban expansion and regional demographic shifts that followed industry's move south and west. Big cities became emblematic in the national urban conversation. New York became the face of "the crisis in our cities." Detroit and Chicago carried the symbolic weight of deindustrialization. Phoenix, Atlanta, Houston, and San Diego were the poster children for the rise of the Sun Belt. As a result, beginning in the 1980s, a great deal of capital and moral energy went into securing the fortunes of these larger places. In the process, large cities acted even more strongly than they had in the past as magnets for young people from smaller urban areas, unleashing what could be called a "revolt from the small city."

Meanwhile, federal urban leadership had fallen by the wayside as free market ideology in the 1980s began facing down government involvement of any kind in economic affairs. Indeed, it has been three and a half decades since the United States has had anything even approaching a coherent federal urban policy—that is, since President Jimmy Carter's short-lived National Urban Policy initiatives. During the interregnum, private developers filled the void. They already had huge subsidies dating from the 1950s, such as federal accelerated-depreciation tax incentives and local tax breaks that encouraged shoddy, temporary construction on open land, or "greenfields." When cities tried to contain the resulting sprawl (earliest and most successfully in Portland, Oregon, which in 1974 drew up an urban growth boundary), they eventually found themselves up against a vociferous property rights movement consisting of both farmers and property investors eager to cash in on the bounty created by their land's proximity to cities. Absent federal direction—funding for neighborhood-based "enterprise zones" was about the only counter-sprawl tool made available to cities in the 1980s—urban affairs were left to an array of federal agencies executing unintegrated policies; private foundations and think tanks such as the Brookings Institution and the Urban Institute; professional associations of mayors, chambers of commerce, and state and local officials; and private developers. The establishment of the U.S. Department of Housing and Urban Development (HUD) in 1965, which concentrated on affordable housing, had never distinguished between large and smaller cities: its short-lived Office of Small Cities, whose authority was transferred to the states in 1982, covered settlements of 50,000 or fewer. Nonetheless, smaller cities felt the brunt when the Reagan administration slashed the budget for HUD while implementing a series of tax reforms that encouraged private commercial investment, which was disproportionately funneled into big metro downtown projects and exurban development. Also working against the interests of smaller cities was the deregulation of the banking system, which removed regulatory power from the states and had the effect of disempowering local banking institutions that once had a hand in small-scale urban development. Smaller cities got lost in this morass of competing interests and big-money constituencies; metropolis was king in a universe where the suburb had become God.[25]

Smaller cities became further marginalized imaginatively, along with anything of modest ambition, by the general cult of gigantism that has marked economic bubble culture since the mid-1990s. The mood of triumphant growth, inevitable and unlimited and much deserved, was perhaps best captured in a frequently aired beer commercial of the period that featured behemoth twenty somethings looming over the Rockies, casually flirting and tossing around a football: the sky's the limit. A similarly "supersize-me" sensibility could be found in the proliferation of McMansions and ever-larger SUVs. The triumph of "free" market principles masked an unprecedented concentration of power in our economic and banking institutions and, with the George W. Bush administration, in the executive branch of the federal government. Under the cover of "diversity," cosmopolitanism itself became hardened and disfigured by this unsustainable race to the bottom line of ever-expanding market returns.

Closely tied to this obsession with all-things-mega was an ethic of disposability, drawn from ramped-up consumer culture and applied heartlessly to any "loser" that could not stand up to the inexorable forces of privatization and globalization. Smaller industrial cities fell under that wheel, too. Not only were they small and even shrinking, but their pride as home to America's producers—the world's bread basket, the shop-floor heartland—already eroded by decades of deskilling and mechanization, now became a joke. It's fair to say that the 2007 Pixar motion picture *Wall-e*'s portrayal of the denizens of postapocalyptic consumer culture—floating, fat, and feckless—captured something of the cosmopolitan regard for places like Dayton and Peoria.

* * * * * * *

Still, smaller industrial cities are not simply victims, haplessly fielding the consequences of decisions made by their global taskmasters. They've had choices too, and they still do. Countless talented planners and devoted citizens have done yeoman's work to secure the quality of their neighborhoods and revitalize their downtowns, create arts amenities and preserve their historic architecture, clean up their rivers and park systems, build job-creating medical centers and expand institutions of higher learning, improve their public schools, and redress longstanding class and racial inequities. Some, however, have been trapped by their own parochialism. Civic leaders in these places, often older and white and indebted to time-worn patronage systems, have found themselves beset by increasingly impoverished, minority urban populations and faltering manufacturing-based economies—troubles they are ill equipped to manage. Moreover, there is a longstanding perception among many in these communities, not entirely unfounded in the absence of a robust national industrial policy, that environmentalism hobbles industry with regulations that make it hard to do business and thus threatens job creation.[26]

Many political and economic leaders of small industrial cities have barely wrapped their heads around the idea of a "new economy," much less a low-carbon one. As one prominent urban analyst, who focuses on the Midwest, put it to me, "I continue to be utterly bewildered at the decisions small cities make, and the unwillingness to change their focus away from traditional manufacturing to new industries, even when old manufacturing has been totally destroyed."[27]

These places have been battered, disproportionately fractured by urban highway development and the flight of retail to the suburbs; eviscerated by deindustrialization and the global economy; at war with "environmentalists" perceived as threatening their manufacturing and agricultural economic self-interest; and ignored or mocked by the cosmopolitan culture of the global city. In fact, the culture of the new economy, with its embrace of "smart" knowledge workers and the creative class, and more than a few of the urban theorists who call for the amenities to attract them, have openly disdained the dignity and talent of working people. Clearly a cultural change must accompany any meaningful and effective transition to a low-carbon economy, and it must go both ways. The practical political appeal of the emerging green economy, if we do it right and quickly, is that it appeals to blue-collar self-interest with well-paying green-collar jobs: in this scenario, the new economy must be integrated into a new version of the old economy, secured by a manufacturing base making and maintaining things that people need in a low-carbon world—railway cars, wind turbines, solar panels, weatherizing materials, and the like—as well as by a more localized approach to the cultivation of food.

Smaller cities in other parts of the world are already planning for a post-oil future and have a big economic jump on the United States in doing so. As of 2010, according to the United Nations, more than half of the world's population lives in urban areas, with small cities of between 100,000 and 500,000 absorbing most of the growth—and mostly in Asia. China alone is home to 25 percent of the world's 961 small cities, and both numbers are expected to grow dramatically.[28] With millions of rural dwellers flooding into the east coast megacities of Shanghai and Beijing, which already number more than 10 million people each, the government in 2001 announced its ambition to build 400 small, sustainable cities of some half-million people to house the anticipated 400 million rural-to-urban migrants by 2020. That hasn't quite come to pass, in spite of the whetted appetites of Western ecodesigners eager to cash in.[29] Nonetheless, China is quietly building smaller low-carbon cities on its own terms—Turpan, on the old western Silk Road, is something of a model—and constructing the transportation infrastructure to connect them.[30] Sweden, for its part, has since the 1980s been building a network of ecomunicipalities, which now comprise 25 percent of the country's settlements—a model that has been replicated throughout Scandinavia and has recently made its way to the United States. Of course, a few smaller American cities have been devoted to low-carbon principles for some time—the university towns of Austin, Texas, and Madison, Wisconsin, come to mind—and communities of various scales have embraced the Swedish idea. Smaller industrial cities, however, have been generally slow to see its value.[31]

This is an opportune moment of great urgency for all American cities, large and small. In Barack Obama, we have our first urban president since Franklin D. Roosevelt. Within a month of his inauguration in January 2009 and beset by the worst economic crisis since the Great Depression, Obama created the White House Office of Urban Affairs. Soon after he named a special advisor for green jobs. By June, his secretaries for the Department of Housing and Urban Development, the Department of Transportation, and the Environmental Protection Agency announced an unusual joint effort, an interagency Partnership for Sustainable Communities, intended to break through the budgetary siloing that blocks thoughtful, efficient interagency programming. Yet at the time of this writing, low-carbon federal initiatives have been stymied by a deadlocked Congress. In view of global warming, that is a tragedy of world-historical, even existential proportions.

Consider what some have tried to get on the table. The Community Regeneration, Sustainability, and Innovation Act would offer opportunities to reshape older industrial cities that have been gutted by job loss, depopulation, and decades of inequitable funding for suburban sprawl. A climate change bill would provide a regulatory framework for reducing greenhouse gas emissions and incentives for the development of renewable energy systems. The transportation reauthorization bill—which comes up every six years and, under ordinary circumstances, would have been passed by now—could swing our national budget priorities toward better public transit. Indeed, the Obama administration's 2009 stimulus package and early federal budgets together signaled a desire for structural shifts in urban, energy, transportation, and, to a lesser extent, agricultural policy. In the face of deepening political polarization, however—formalized by the success of Republican Tea Party candidates in the 2010 midterm elections—low-carbon, urban-centered policy initiatives have had to take a piecemeal approach while an ever more reactionary right wing succeeds rhetorically in reducing "government" to "socialism."[32]

* * * * * * *

As I was preparing this book, I ran across a small notice in my hometown newspaper, the *Baldwinsville Gazette and Farmers' Journal*. Published just weeks before President Kennedy's assassination and two years before the paper folded, it announced that my sister and I had won first prize in a Halloween costume contest. I remember that my father had created our get-ups out of cardboard and paint: we were Heckle and Jeckle, the cartoon magpies from a popular TV show that middle-class kids across the country watched every Saturday morning while their parents tried to catch a break. Featured on the same page are three big ads for cars and snow tires and a thank-you notice from the Republican Party to voters who had supported its winning candidates. A long story on the retirement of the local fire chief, who had earlier worked in one of the village's numerous machine shops, concludes with an announcement of his testimonial dinner, stating matter-of-factly, "It will be a stag affair."

It's hard not to wonder what might have become of places like Baldwinsville, which shared the cultural biases of its time, had all those cars outfitted with snow tires not been provided with expensive new highways that paved over the town's farmland and got people to Syracuse just a little faster. And for what? Those highways eventually crisscrossed the entire metro area, drawing people away from the city to suburban office parks, housing developments, and shopping centers that sucked retail out of downtown, leaving the people of Syracuse to fester in poverty, further deepening the country's tragic racial divide. Is it a fixed historical axiom that the farther settlements are from urban centers, the more close-minded their people are? That the bigger the city, the more "modern" and thus more civilized it inevitably becomes? And that, therefore, we shouldn't cry too many tears about smaller industrial cities' demise?

Well, no. Smaller cities could be placed at the center of a relocalized low-carbon world in a truly new urbanism. The technology is under development, the private capital is poised for investment, the foundation world is paying attention, urban advocates are showing greater sensitivity, and the Obama administration might just have the moral and political imagination to advance a sustainable future—one intimately tied with cities of all sizes. The question now is whether smaller industrial cities can themselves make the case that their destiny is critical to that future.

Small, Gritty, and Green

Worcester, Massachusetts. S. Salina near Lafayette, circa 1938. Reprinted by permission of the Onondaga Historical Association.

1

Against "Shapeless Giantism"

Self-described "environmental Nazi" Julie Backenkeller is preoccupied with an issue New Yorkers don't have to think much about: whether to put dead deer in the city landfill.[1] In 2002, the dreaded chronic wasting disease that had been imperiling deer herds in the western states for several decades hit Wisconsin hard. The only way to control the always fatal neurological ailment is to kill the afflicted creatures, but there's a hitch: the disease is spread through the ingestion of prions, rogue proteins that are responsible for such disorders as the better-known mad cow disease. The best way to destroy prions is through an intensive, and expensive, incineration process. To Julie's horror, the cash-strapped city of Janesville proposed disposing of the deer carcasses in its landfill, located in an old sand and gravel mine. Julie feared that the deadly prions might leach into the city's water supply, endangering the health of her two little boys. Soon after we met in November 2009, the city won that fight, but it only emboldened her to push harder on other fronts.

Julie's education in urban politics and environmental protection began one day in the mid-1990s while sunning herself next to a Janesville public pool and keeping an eye on her kids. She struck up a casual conversation with a woman sitting next to her, who turned out to be a reporter with the *Janesville Gazette* working on a story about the city's plans to expand into the rich farmland to its northeast. Before long, the controversy had unleashed a human tornado in the person of Julie Backenkeller. A former beautician who today runs a business-to-business courier service with her husband, Julie is warm, quick-witted, and passionate—her salon clients must have loved her—with a droll, self-deprecating, we-are-so-doomed midwestern sense of humor. When she's really steamed, she uses the F word with abandon. Julie, who is in her early forties, is a Janesville native whose family goes back five generations in this city of some 60,000 just over the south-central Wisconsin border. She's still mad that her family lost the small store it ran in her old neighborhood when the town shut down commercial zoning in the area. "This is *my* town," she says, and she doesn't care that the city's leadership is sometimes put off by her "disrespectful" demeanor—which she alludes to with a hint of smart-alecky pride. Over the past seventeen years, she has launched many forays into environmental civic activism, developing recycling programs in her sons' schools, helping to create a city council sustainability committee, writing a green column for the *Janesville Gazette*, and in 2009 running as a last-minute write-in candidate for a city council seat "to call the other candidates on their crap." And she hasn't let up for one minute on that most intractable environmental problem of all, the one that everywhere bedevils sustainability advocates: suburban sprawl.

I had decided to visit Janesville for one primary reason: GM had just shut down its Janesville assembly plant as part of the auto industry's desperate effort to reinvent itself with the 2009 federal bailout. It's not hard to see why the facility was closed: it produced gargantuan SUVs—Chevy Tahoes, Suburbans, and Yukons—and high fuel prices were shrinking that market. Since 1919 Janesville had been home to GM's longest-operating assembly plant, and I was interested in learning how the city, with its longstanding, formative relationship with the auto industry, was coming to grips with the very forces that led to the plant's closure. There were signs of real potential here: just before GM announced the plant's shuttering, the city council designated Janesville an "ecomunicipality," based on an urban-based sustainability model developed in Sweden and embraced by 1000 Friends of Wisconsin, the state's preeminent environmentalist sprawl-busting organization.[2] As part of that pledge, it formed a sustainability committee of interested citizens charged with advising the city on low-carbon initiatives. When I arrived in Janesville, then, I stumbled on a big fight over land use that was only tangentially related to GM and the loss of its 2,400 remaining jobs. The economic slowdown that laid waste to GM also held up a slew of development projects, giving Janesville residents time to fashion thoughtful long-term plans to curb sprawl.[3]

Janesville is an attractive little city nestled along the banks of the Rock River, which provided power for its earlier industries. The city was formally designed in 1920 by famed landscape architect and city planner John Nolen, who was hired by the local chamber of commerce as GM settled in. The Parker Pen Company was also a major employer until 1987. Nolen—who brought the City Beautiful aesthetic into his work as one of the country's earliest comprehensive plan makers—is better known in these parts for his design of Madison, forty miles north of Janesville, and for planning Wisconsin's park system. But his sensibility is everywhere in evidence in Janesville, too. Nolen had an acute sense of "fitness and appropriateness," by which he meant designing to scale for local context, attentive to balance between formality and informal "nature," elegant vistas and calming green landscapes. He wrote of the need to instill "love and pride in local traditions and local ideals," which, in the prairie of lower Wisconsin, meant building out from the river and integrating parkland with an architectural dignity suggestive of prosperous commerce, both agricultural and industrial.[4] To this day, Janesville is known as "Wisconsin's Park Place," and the downtown area retains a semblance of Nolen's casually elegant sense of design, appropriate to a small city in the Midwest. (So, too, does Flint, Michigan, another GM town that Nolen designed.)

Several of Janesville's characteristics set it apart from most other smaller industrial cities. For one thing, its population is overwhelmingly white—more than 95 percent white—and has been from the beginning. When GM converted to war production in the 1940s, company management told black migrants who came to work in the factory to live in Beloit, fifteen miles south, where a marine engine plant had already drawn black workers during World War I. Although many postwar middle-class Janesville residents were drawn to the quarter-acre lots of suburbia, the city did not undergo the dramatic white flight that generally afflicted older industrial cities in the 1960s. As a result, its public school system is relatively intact, and sprawl, until recently, has been less intense than elsewhere.

Also unusual for an older industrial city, Janesville's population has been growing steadily since the 1920s, and it doubled between 1950 and 1970—the very period when most of these places began to shrink.[5] And there's good reason to think that in spite GM's troubles, Janesville could continue to grow. Over the previous ten years, as the local assembly plant began to pare down, city leaders diversified Janesville's economy with a large medical center, and another is in the works. It has a large Seneca Foods processing facility that provides a market for the vegetable produce grown in the surrounding agricultural economy. And increasingly it is attracting commuters to Madison, 40 miles north, and Rockford, 38 miles south. Interstate 90, which connects these cities (along with Beloit), has been identified by the city's planning consultant as a "growth corridor."[6] The question here is, What kind of growth, appropriate to its size and natural assets, will prepare it for a low-carbon future?

When I decided to visit Janesville in fall 2009, I contacted community development director Brad Cantrell, who had ushered through the city's new twenty-year comprehensive plan. Just passed by the city council that spring, it was intended to replace a plan established in the 1980s. The earlier plan had permitted open-land development of an enormous big-box center, Pine Tree Plaza, and nearby residential communities on the city's northeast side, just off I-90—development that had gotten Julie Backenkeller's back up that day by the pool. Janesville's new plan was drafted in response to a pioneering state law requiring localities to craft comprehensive plans along smart growth principles by 2010. The smart growth movement, which is concerned with limiting land use practices that facilitate sprawl, seeks to transcend the polarizing "development–no development" debates that had gripped so many communities since the 1980s. Its more intentional approach to development, which includes greater community participation, seeks to balance the need for jobs and economic development with the preservation of ecosystem integrity and open spaces. To that end, it promotes respect for established neighborhoods, pedestrian-friendly streetscapes, and variety in design—of housing types, density levels, and transportation choices—and it encourages transit-oriented development and mixed residential and commercial zoning. In other words, smart growth aims to arrest the frenzied thoughtlessness and uniformity of sprawl, cultivate a sense of place, and provide alternatives to the auto dependency that is sprawl's lifeblood.[7]

Brad Cantrell is a reserved man in his fifties who has worked in the city's development office for twenty years. He is also a member of the local Metropolitan Planning Organization, which, like similar organizations across the country, acts by default as the area's only regional planning authority: it funnels federal dollars into state and regional transportation projects. He seems vaguely oppressed, as though burdened by the compromises bureaucrats invariably shoulder.

I met with Brad and planning services manager Duane Cherek in a conference room in the Janesville Municipal Building—a blocky 1970s-era affair surrounded by parking lots. We talked about Janesville's recent economic history and John Nolen's original plans for growth as reflected in its large-capacity sewer and reservoir systems. Since I was especially interested in what the city was doing to become more sustainable, we discussed stormwater drainage systems and plans for spending federal stimulus money on weatherizing buildings and upgrading the electrical infrastructure with light-emitting diode (LED) lighting. We also talked about those deer carcasses and the city's landfill program, which brings in revenue and will be turned into parkland eventually. Then they unfurled enormous maps of the sort planners use, marking out the city's comprehensive plan.

As I narrowed my eyes and viewed the maps in fuzzy abstraction, Janesville looked like a giant Pac-Man head—a roughly circular shape with the lower southeast quadrant cut out. "The city can grow in any direction," Brad explained, "since it doesn't have natural boundaries or nearby municipal competition." The city's development has been "lopsided," he continued, favoring growth on the east side near I-90. The new plan encourages development on the west side to right the balance. Besides, he said, "Janesville is surrounded by some of the best farmland in the world—especially on the east side—and the future of the local economy will be ag based, not manufacturing based." The plan expands on the rural-suburban development that already exists, he pointed out, while encouraging 15 percent higher density in the city.

We met for an hour, and then Brad gave me a long tour of the city—an unplanned act of generosity. When I asked him about the role of the Sustainability Committee, he observed that "it was just beginning to define its purpose" and seemed primarily interested in green-building principles, water use, and energy efficiency. We talked about the school system and medical center (the city's two largest employers), as well as the city's desire to buy the enormous GM plant when the company puts it up for sale. As far as Brad was concerned, the loss of GM was a mixed blessing since its union wage scales "may have kept the city back." We also shared delight in the design bones laid down by John Nolen and the beauty of the city's residential architecture and old commercial buildings. Brad pointed out that the popular downtown hockey rink was about to be moved to a new facility to make way for a larger fire station in the city's center. And then we drove to the edge of town, and I saw the hockey rink's new site on city-owned property. "It's on greenfields," I noted quizzically. And it was on the southeast side. "This is where young families live," he replied. Farther down the road, we drove past "some weird bird farm," as I had written in my notes just hours before while driving into Janesville. It stretched over some sixty acres. "MacFarlane Pheasants, free-range smoked and dressed," read the sign as I had rounded the gigantic corner lot. "What's going to happen to this?" I asked Brad. "Oh, they're moving elsewhere," he replied. "It's slated for development. They'll be fine. It's a very successful business, and the owner was happy to sell the land at a high price."

When I first learned of Janesville's sustainability committee, I had contacted a prolific, seemingly sensible local blogger to ask whom I might talk with about it. Without hesitation, he referred me to Julie Backenkeller, a member of the committee and cofounder of the Rock Environmental Network. "You have to meet with me," she wrote back emphatically. It was the first of an onslaught of e-mails she fired off to bring me up to date. Before long I found myself in her kitchen, with two other citizen activists, eating the best homemade pumpkin soup I'd ever dreamed of tasting. Julie had insisted on feeding me, an unfussy eater who had subsisted on fast food while on the road, and finally broke through my demurrals.

As it turns out, Brad Cantrell is Julie's nemesis, and she's not too happy with Duane Cherek either. She spent three hours with me elaborating on claims she had outlined in e-mails, supplemented by soil maps and photographs of land planned for development. The Rock Prairie, on which Janesville sits and which extends to the east of town, consists of Plano silt—the richest farmland on earth. Scattered throughout the Upper Midwest, it's found in abundance in only two other places: east of the Ural Mountains in Russia and west of Rio de Janeiro in Brazil. "That's it!" Julie had written. "Janesville's Comp plan, which adds 10,000 acres of prime farmland for development, in the face of economic disaster, was hastily adopted last March [2009], by an uneducated city council," she continued. "Our city ordinances require two public readings before the adoption of the document, and in a sneaky-bastard, last-minute addition to a council agenda, Cantrell requested that the city council 'suspend' the ordinance, so that the public readings were not required. CAN YOU BELIEVE THAT??? I absolutely came unglued! Our Sustainability Committee had requested the chance to review and make suggestions on the document, but Administration did everything within their power to get that document passed before we could get our teeth into it." She further charged that "there are three wealthy (slimy) developers that have bought up land on the prairie specifically for development" and that the city was "trying to sneak through all kinds of infrastructure to areas that are not even ready to be developed, just to have an excuse to develop them." Meanwhile, she continued, "our inner-city infrastructure is crumbling, unemployment and home foreclosures are at an all-time high, and bankruptcies have hit the 30 percent mark."[8]

Julie handed me fellow Sustainability Committee member Alex Cunningham's preliminary comments on the draft documents that would become the city's comprehensive plan—comments that she claims the city council never responded to. "We must conclude," Cunningham wrote, "that these documents do not constitute a plan" but "a vision for uncontrolled real estate speculation" on the outer edge of the city. Among Cunningham's criticisms was the charge that the plan used inadequate methods for forecasting future population growth (putting it at 80 percent over that projected by the state) and that it omitted basic requirements of the state's smart growth legislation, such as inventorying the city's vacant land and buildings and forecasting employment. As a result, it "allocates virtually *all* future growth to new lands," disregarding tax equity issues and meaningful farmland protection. "*Redevelopment of the downtown and commercial corridors and existing industrial properties will be seriously inhibited*," he stressed, "by allocating some 13 million square feet of commercial uses and extensive greenfield industrial opportunities at the urban fringe, *while only affording redevelopment of a few dozen parcels in the 'tired' commercial corridors and downtown area which is largely slated for preservation*." Here was, it would seem, a robust plan for developer-driven sprawl of the sort that has beset cities large and small for more than fifty years.

It's important to note that Brad Cantrell too is critical of sprawl, and he's fond of downtown Janesville. He sat on a state transportation committee planning for a highway bypass to reduce traffic in the city and argued for not running it too far out because it would "promote sprawl and development."[9] On our tour, he told me that the city had not ripped out its rail tracks in the event that it might one day want to build light rail. He noted that the city might one day purchase the privately owned Centerway Dam, a spillover hydroelectric dam in the middle of town. He was also careful to point out that the city has extraterritorial jurisdiction over land use 3 miles from the city's limits, and Henke Road, a mile to the east of Highway 14, the city's current eastern border, was therefore a logical boundary for building outward, as reflected in the comprehensive plan. "Logical Boundary???" Julie had scribbled on the backs of four photographs of the disputed area showing tilled farmland from all directions at a crucial intersection of the road.

Planning for What?

The worldview reflected in Janesville's comprehensive plan is consistent with an emerging, influential vision for the Midwest spelled out by Richard C. Longworth in his 2008 book *Caught in the Middle: America's Heartland in the Age of Globalism*. An Iowa native and veteran Chicago journalist, Longworth speaks in a tone of seasoned, take-no-prisoners realism worthy of Raymond Chandler. His message to the Midwest is that it's time to man up and face facts: globalization is only going to intensify and those comfy middle-class union-scale manufacturing jobs and small farms are history. The economic future of the United States, he argues, belongs to knowledge industries, and the idea-generating, educated creative classes—open-minded professionals, artists, gay men, "bohemians"—that fuel their growth. One of the peculiarities of the creative classes, he observes, echoing Richard Florida, is that they like to be near each other, to have face-to-face contact where they can "bounce ideas around." In this reading, a city's success will be measured not by its size but by its ability to attract a critical mass of such people who can generate new ideas for export in the global marketplace. Small university-based cities such as Madison and Ann Arbor, for example, hold just as much appeal for "talent" as Chicago and New York do.

Globalization is ruthlessly efficient, Longworth observes, resulting in geographical compactness that accommodates the creative class's "need" to hang together; it is also desirable to venture capitalists who want to be close to their investments. This drive toward concentration is facilitating the rise of megaregions anchored by big cities and including smaller communities "smart" enough to integrate with them. Globalization is no respecter of political boundaries, and Longworth's larger argument is that the Midwest must muscle through now-antiquated jurisdictions that pit states, universities, and cities themselves against one another in a race to attract investment—what has been called "smokestack chasing." Moreover, Longworth argues, the old nineteenth-century political boundaries can do nothing to avert the emerging "rural-urban split," since small rural economies in the Midwest have long been dependent on what has remained of manufacturing, which is in its final death throes. "To live in the hinterland of a nonglobal city," he writes in a characteristic spirit of tough love, "is to be condemned not necessarily to poverty. Just irrelevance."[10]

The Janesville Comprehensive Plan is farsighted from this point of view. It seeks to situate the city within a megaregion that extends from Chicago to Minneapolis/St. Paul, while taking advantage of its proximity to Madison—a creative-class magnet. Its architects have good reason to plan for growth and to do everything in their power to blow through the political obstacles in their path—obstacles framed by an older, unenlightened way of thinking. Future generations will thank people like Brad Cantrell for seeing to it that Janesville did not become one of globalization's "losers"—a word Longworth uses often to chide his fellow midwesterners.

But what if Longworth is wrong about globalization? He wrote his book (and Brad drew up Janesville's plan) at the peak of a ten-year housing bubble, one of three successive speculative bubbles to make their way through the American economy since the mid-1990s. Together they created an artificial sense that we had entered a new era of permanent market expansion with near freakish capacity for growth. Yet the financialization of the economy, a crucial aspect of globalization as Longworth conceives it and whose "creative debt instruments" he praises, led to the worst recession since the Great Depression. It remains to be seen how this correction will shake out. Besides, globalization relies on cheap long-distance transportation and industrial food production, both highly dependent on finite reserves of petroleum, which are susceptible to spiking fuel prices and mounting alarm about climate change. Catastrophic weather and war are also disrupting the triumphant arc of unlimited market growth. All of these chinks in the global supply chain have led even mainstream institutions to start planning for energy and food security as a matter of course. It will surely have an effect on the usual ways we've come to talk about globalization, including its disregard of political jurisdictions.

Although not everyone would agree that, in Longworth's words, "the role of the states is to get out of the way," many decry the Midwest's civil township system. An extra layer of governance created early on by the counties, townships can be annexed in pieces by growing municipalities that have "extraterritorial jurisdiction" (in Wisconsin, 3 miles from a city's borders) and where urban proximity drives up land values, leaving landowners eager to sell.[11] Hence, Janesville's ambitions for the "urban reserve," which lies within several townships. And yet the one person who has won the respect of all parties in the dispute over Janesville's comprehensive plan is LaPrairie Township chairman Mike Saunders. LaPrairie, whose population stands at 924, is adjacent to the southeastern border of Janesville, and is home to a 5-mile stretch of I-90/39. With that enviable highway access and the sprawl pressing in from the north, in Harmony Township, developers are pushing hard to get into LaPrairie. LaPrairie's resistance to that pressure, led shrewdly by Mike and his predecessors, accounts for the mouth of Janesville's Pac-Man-head shape.

Although almost everyone who lives in LaPrairie is a farmer, Mike is not. He knows a lot about farming and how farmers think, however, since he spent years as an agricultural insurance adjuster poring through farmers' books. On his property he has just a few pear and peach trees that he's breeding for seeds and six or so sheep that graze on the hill leading up to the modest frame house he shares with his wife, Barbara, of forty-two years. A cultivated self-described "California girl" who tires of the prairie every so often, she's persuaded Mike to pay a two-week visit to Paris with her the following month. Mike is also well read and worldly, "a renaissance sharecropper," he says, who uses words like *solipsism* yet doesn't take himself too seriously. His favorite vacation spot is in West Fargo, North Dakota, where he takes an annual fishing trip with four buddies.

At the urging of several people, I found myself sitting in Mike and Barbara's small kitchen on the morning I left Janesville, a little bleary-eyed—I was nearing the end of a five-city trip. Mike's an energetic guy in his mid-sixties, fond of the Socratic method. At a fast clip he would unfurl a map or hand me a chart and ask, "What do you see?" When I'd look up at him helplessly, he would provide the answers. And when he'd describe the ways he found to block infrastructure appealing to developers, he'd sing in a higher register, "I don't *think* so . . ."[12]

When Mike became chair of LaPrairie in 2003, after serving on the board for six years, he set about using every available legal mechanism to preserve the town's land for farming. When the nearby Seneca Foods processing plant on the border with Janesville expanded in 2005, he made an arrangement with the city. Instead of overwhelming the city's sewer system with the washwater used to process vegetables, Seneca now diverts it to LaPrairie's farmers, who use it to irrigate their crops. Should this arrangement change, the city's sewerage charge would increase. Janesville thus has a stake in maintaining the plan—and in keeping LaPrairie in farmland. In 2005, he also participated in updating the county's agricultural preservation plan, first developed in 1979. Later, in 2007, he put together a smart growth comprehensive plan (rare for townships), for which he enthusiastically solicited the required "citizen input" and used it to cultivate conversation, debate, and consensus. Mike pointed out to me that only six or seven counties in Wisconsin grant local control of zoning, and LaPrairie sits in one of them. The town has used that power to draft zoning ordinances inhospitable to development, which are of particular value in what state law calls urban agricultural transition areas—undeveloped land adjacent to a city of no less than 35 contiguous acres. These areas, which ring most smaller cities in the agricultural Midwest, are expected to undergo rapid, dense development if population increases. It is to these most threatened areas, bordering Janesville, that Mike and his allies have devoted much of their effort. The Janesville plan tellingly calls its own adjacent farmland border "urban reserve."

One of the most common and effective ways that sprawl wedges into farmland is through popcorn development. Here, developers buy up relatively small, scattered properties and work with the city or county to prepare them with sewer and utility lines and improved roadways for anticipated building. Scattered development has a cancer-like effect. "Once non-agricultural uses are located near farmland, the farmland may be assessed at a greater value," notes the LaPrairie comprehensive plan, "causing the farmers' taxes to increase. Also, when farmers foresee possible speculation on nearby land for development, they tend to invest less in their farming operations, an effect known as 'impermanence syndrome.'"[13] Making matters even more difficult, Mike told me, is that efforts to put agricultural land into trust through nonprofits such as the Nature Conservancy or through state-funded programs all too often can't compete with the price that developers can offer for the land. And then there are tax-increment financing (TIF) programs that have been used, especially in smaller cities over the past forty years, in place of reduced federal and state funding for public development projects. When a parcel is declared a TIF district, it becomes eligible for deferred taxation that won't go into effect until the project is complete; the designation can also be used to borrow against. Intended primarily for blighted urban areas, TIF financing has become increasingly popular in already development-prone urban fringes—a sweetener to the developer's pot for which the program was never intended; it too contributes to popcorn development.

In effect, Mike Saunders and the farmers of LaPrairie have devised ways of blocking popcorn development. After putting together their comprehensive plan, they immediately spent a year and a half getting their zoning ordinances and deed restrictions "up to speed," as Mike put it to me—a process that often lags many years behind the creation of a comprehensive plan. Local zoning was already strict and conformed with Wisconsin's Exclusive Agricultural Zoning guidelines, requiring minimum lot sizes of 35 acres in the land most endangered by sprawl and permitting only agricultural structures and improvements on that land. It also imposed low-density requirements. Nonetheless, between 1999 and 2006, LaPrairie lost some 47 acres to Janesville; between 1982 and 2002, the county as a whole lost 18,000 acres of farmland, mainly through annexation. So Mike got creative in ways he recounted with pride. He and his cohorts established density requirements that put a ceiling on the number of housing units allowed on a set number of acres of land. Few residential developers would be enticed by that. They severely limited driveway lengths and sign sizes—no McMansions or big stores here. They began charging for existing signage and went to court to ensure that signs along the section of I-90/39 that runs through LaPrairie must be at least 4,000 feet apart, discouraging the crowded signage retailers use to vie for drivers' attention. And for the coup de grace, they implemented stringent telecommunications tower height requirements—indispensable infrastructure for commercial and residential development.

With all these changes, Mike told me "the economics kicked in," by which he meant that the city of Janesville and the county could see more clearly that agriculture could be an economic driver for the region, in what he called a "corporate-partner structure." The area had always done well with commodity crops, dating back to the days when Janesville was a major sugar processing center during World War II and LaPrairie's soil produced 18 tons of sugar beets per acre. Today most LaPrairie farmers grow corn, with crop yields of 11 ton, or 250 bushels, amounting to $1,500 per acre; some others grow soy, yielding $650 per acre. The town had recently become home to three major seed testing stations, for Pioneer, Monsanto, and Syngenta. Although commodity agriculture is still big in LaPrairie, the township also raises a substantial mint crop that supplies a pleasant zing to the early summer air and grows vegetables for the Seneca plant. Not all of LaPrairie's farmers are natives, and quite a few farm to supplement their incomes as white-collar professionals, which brings the average annual income to a relatively high $60,000. In the historic 2008 presidential election, their vote split exactly in half between Democratic and Republican Party candidates. All in all, the small agricultural township of LaPrairie does very well and brooks its differences through a common interest in the land.

As I prepared to leave, Mike handed me a pear. I asked him if he sells his fruit or his sheep. He does sell the sheep, he said, but it's informal—"people just know"—and he doesn't have to spend time marketing them. His responsibility to the preservation-of-the-land ethic, he said, was "first and foremost." I then asked him if he had read Michael Pollan's *The Omnivore's Dilemma*, which is critical of the commodity-based monoculture that so many area farmers practice and which the town was supporting through its seed testing stations. He had. "What did you think?" I asked him. "I honestly don't know," he replied thoughtfully. "It's a debate worth having, that's for sure."

Julie, meanwhile, got herself appointed to the local committee of the state's PACE (Purchase of Agricultural Easements) program, which went into effect in 2010 and funnels matching funds into local programs that buy development rights from farmers, guaranteeing the land's permanent agricultural use. The City of Janesville hasn't put up much money to participate—only $700,000—but she hopes that it's another step toward shaping development in a sustainable way. The program's long-term intent is to encourage urban growth boundaries that provide both developers and farmers a sense of where development makes the most sense, and to plan slowly and carefully for it.[14]

The Janesville metro area is hardly a hotbed of alternative culture: its current U.S. representative is conservative Republican Paul Ryan, and although it is gifted with prime soil on all sides, little of it is used to produce fresh food for the local population. The city supports just one community-supported agriculture (CSA) farm subscription program and one store that sells organic produce. But the fight over preserving its fertile soil for agriculture is preparing the area for local sustenance in a low-carbon future. Although the process has been messy and conducted very much in terms of immediate economic self-interest, Janesville may be poised to one day become a model of sustainability if it chooses to do so. But the old GM town could not possibly do it alone. It also requires substantial shifts in both policy and culture—in the way we think about regional development, land investment, agriculture, and urban scale.

"Spatial Democracy," the Metropolis, and the Common Wealth

The tumult in Janesville and elsewhere over curbing sprawl, preserving the surrounding farmland and ecosystem, and framing civic identity within the broad urban region resurrects older debates about the place of small industrial cities—or lack thereof—in planning for urban settlements. Revisiting that history is useful, even essential, to charting out a vision of what these places might become and understanding how they could easily get lost in today's gathering consensus about global megaregions.

It helps to remember how new and puzzling industrial cities were when they first took shape in the eighteenth century. Across the world and through time, human beings had amassed themselves in cities for three primary reasons: to create sacred space, mount basic security, and provide a place of exchange for commercial markets. The hinterland was given over to agriculture, which, until the late twentieth century, employed the labors of the vast majority of the world's people. The rise of industrial manufacturing broke this pattern in two ways, unleashing developments that were without historical precedent. Large cities suddenly became much larger at rates never before seen, bulging with both impoverished factory workers and those who lived off the fat of their labors, from bankers, business owners, and commercial traders to shopkeepers, salaried workers, and civil servants. Cities had risen and fallen in size and power in the past, to be sure. But this was different: the industrial city was so polluted, pestilential, crime ridden, and dense, and the land pressures so sudden and so great, that by the 1840s, the emerging middle classes escaped when they could to settle in the less expensive adjacent countryside and commuted to the city for work, an arrangement that anticipated modern suburbia. As a result, modern cities began to lose their moral cohesion, their ability to command civic loyalty to a particular urban place. Meanwhile, the rural hinterland faced radical population loss—a trend that continues apace today, from Iowa to central and western China.[15]

Industrialization also spurred the growth of small cities that owed their existence primarily to resource extraction and manufacturing. At the height of the industrial revolution in England in the early nineteenth century, the cities that experienced the most rapid growth were in Lancashire, long a backwater of rural poverty. Its principal city, Manchester, grew by some 600 percent over ninety years, reaching a population of 540,000 by century's end, and smaller cities in the industrializing region grew even faster.[16] In the United States, "Lancashire" lay in the Northeast and Midwest. By 1850, Lowell, Massachusetts, the birthplace of American manufacturing with the water-driven textile industry, had become the second largest city in the state, after Boston. With the expansion of the railroads and the rise of heavy industry, the same pattern later emerged in the Midwest, where numerous towns grew into small industrial cities beginning in the 1880s.

Some eighty years later, in 1961, Jane Jacobs published *The Death and Life of Great American Cities*, a book pivotal to the broad public conversation about cities and now a classic work of American social criticism. An activist and journalist who lived in New York's Greenwich Village and who hailed from the small industrial city of Scranton, Pennsylvania, Jacobs dramatically altered the way students of urbanism talk about cities even today. "By 'great' Mrs. Jacobs seems always to mean 'big,'" noted Lewis Mumford dryly in his *New Yorker* review of the book.[17] And, indeed, Jacobs wrote in the introduction to her masterpiece that she was unconcerned with either "little cities" or regional planning—scales of great importance to Mumford. Through Jacobs's influence, both small cities and regionalism soon dropped out of the urban conversation for a long time.[18]

Jacobs wrote *Death and Life* in response to the grotesque overreaching of postwar urban planners who, in the name of urban renewal and planning for population growth, set about decimating New York's neighborhoods. She detailed the effects of urban renewal in other big cities too, most memorably in Boston, whose Italian North End "slum"—at that time slated for clearance, following destruction of the nearby West End—Jacobs viewed as a model urban neighborhood. In response to the postwar housing crisis, urban planners sought to concentrate residential density upward in tall buildings, influenced by the French modernist architect Le Corbusier, situated on "superblocks" that replaced the street grid.

"Towers in a park," scoffed Jacobs, with reference to the small grassy enclosures usually included with these plans and intended to provide residents access to "nature." Slum clearance did little more than increase crime, she argued. Dismantling dense, diverse neighborhoods and warehousing the working poor in high-rise projects replaced vital street life, radically reducing the number of eyes on the street that had kept the old neighborhoods safe. Meanwhile, through the autocratic control of Robert Moses, widely hailed as New York's master builder, federal infrastructure funding funneled into massive highway projects not only destroyed city neighborhoods but facilitated suburban sprawl and urban blight.

Jacobs inspired a nationwide movement of civic activists bent on arresting the assault on the city and its neighborhoods by the "growth machine"—the mayors, developers, unions, bankers, and many planners who stood to gain from urban renewal and auto-dependent suburban development. She herself participated in successful grassroots efforts to halt construction of the Lower Manhattan Expressway (which would have plowed under the SoHo district) and to block traffic through Washington Square Park in Greenwich Village. Jacobs's work gave rise not only to freeway resistance fights in several other large cities but also to the historic preservation movement and to neighborhood-level planning advocacy that engaged local communities in the fates of their cities. Anyone who cherishes urban life owes Jacobs a debt of impassioned gratitude.[19]

It is necessary, however, to revisit some of Jacobs's claims. She laid much of the intellectual responsibility for the dismemberment of urban neighborhoods on the planning profession writ large, and particularly on regional planners. By their own account, "decentrists," as Jacobs preferred to call regionalists, wanted to "decentralize great cities, thin them out, and disperse their enterprises and populations into smaller, separated cities or, better yet, towns"—by which she meant suburbs.[20] There's some truth to this claim. But it would be more accurate to say that the regionalists' original vision of decentrism, and the place of smaller cities within it, proposed a path not taken.

Not surprisingly, it was in Britain, the birthplace of the smoky, congested industrial city, where the earliest efforts were made to plan for the city's strange new relationship with "nature," however amateurishly. Jacobs came down hard on the most influential of these proto-planners, Ebenezer Howard, whose idea of "garden cities," first described in *Tomorrow: A Peaceful Path to Real Reform* (1898), she blamed for the profession's later hostility to cities and obsession with "grass, grass, grass." Hardly a professional himself, Howard was a parliamentary clerk who mused about cities in his spare time and who admired John Ruskin and William Morris's community-and-craft aesthetic of the simple medieval town. Garden cities, as Howard described them, would be self-sustaining urban centers of no more than 32,000 people. Each would consist of a core of public buildings and cultural institutions, surrounded by concentric rings for various other essential activities: first, a circular central park conjoined to a round-about "Crystal Palace" for shopping and the enjoyment of amusements and gardens during inclement weather; residential rings with housing of "varied architecture and design," divided roughly in half by a wide grand boulevard serving as both park and location for public schools and playgrounds; and, yet farther out, a manufacturing and commercial ring, rimmed by a railroad to move goods around the periphery and to external markets. Beyond that, Howard proposed a much less densely settled outer greenbelt for agriculture, dairy, and forestry, connected to the core by six radial thoroughfares running through each zone, giving the whole a pie-like shape. Just as important, garden cities would give rise to yet more small cities when they reached their population limits, separated by green belts and connected by "intermunicipal rail," sometimes acting as clustered satellites around a larger "central city" of 58,000.[21]

Howard's legacy abounds with irony. He viewed the garden city idea as an urban alternative to the sprawling congestion of the large industrial city with its intensive land pressure, and as a means of repopulating the emptying, impoverished countryside in a "healthy, natural and economic combination of town and country life." Crucial to Howard's vision, too, was its financing and plans for local self-government: established on inexpensive land purchased by a limited-dividend company, rising rents would recompense investors while the "unearned increment" (or "collectively earned increment," as Howard preferred to describe it) would pay for city services. Howard's ideas have been derided as hopelessly utopian, and yet many years later, he is slammed for inspiring one of the twentieth century's most common forms: the sterile, single-use planned communities that eventually proliferated around cities in suburban rings, particularly in the postwar United States. This criticism is hardly fair or accurate. Indeed, none of Howard's vision for garden cities—their start-up self-financing, their limited growth, their productive capacity in agriculture and manufacturing—survived in the twentieth-century suburb, aside from their intentional nature as planned communities. Yet Jacobs rooted the hubris of all modern planning in Howard's idealism, conceived in complete ignorance, she argued, of how cities actually work.[22]

Howard had an enormous influence on Lewis Mumford and the circle of regionalists (organized as the Regional Planning Association of America) with whom he associated in the 1920s and 1930s. Mumford learned of Howard's work through his mentor, Scottish biologist Patrick Geddes. Geddes expanded the frame of reference for the garden city to the region considered as an ecological whole. Writing on the eve of World War I, Geddes argued that urban settlements should be integrated economically and socially with their natural endowments and be attentive to the organic limitations of their growth. To that end, he pioneered the practice of taking careful, comprehensive surveys of human and natural geography—hydrology, soil, topography, and climate studies—as a necessary prelude to the planning process. Geddes's ecological regionalism also had the capacity to address the character and needs of communities of various scales.

Central to the American regionalists' vision is a critical appraisal of the metropolis, most thoughtfully (and polemically) articulated by Lewis Mumford. A native New Yorker born in Queens in 1895, Mumford, like Jacobs, was an intellectual of a sort rarely seen today: a generalist who didn't graduate from college and who wrote original criticism for nonspecialists—the general reader. He was hardly "anti-urban," as Jacobs charged.

Mumford shared his urban reform contemporaries' concerns about congestion, poverty, and high rents, but he was just as troubled by New York's monopolization of culture and control of the national economy, which he spelled out in a 1922 essay on New York called simply "The City." While New York had built for itself grand institutions worthy of true metropolitanism, he argued, New York financiers exported its chain stores and advertising and Coney Islands and movie houses to the rest of the country, where "spiritual masturbation" and "zaniacal optimism" substituted for authentic local culture. Mumford bemoaned the "inchoate" state of American industrial cities west of the Alleghenies, which owed their existence exclusively to industry and whose "chief boast" was their "prospective size." In place of a genuine civilization, these cities were sprouting a derivative metropolitanism, a "pseudo-national culture" that "mechanically emanates from New York," from their architecture to their manners. Here, what passed for culture was confined to shopping or amusements on one of the innumerable Broadways that had cropped up in each one of these "submetropolises," in slavish imitation of New York's famous thoroughfare. New York had amassed the fortune that paid for its opera houses and museums—built during the nineteenth-century depressions that took their highest toll on the heartland—because it owned the land and productive machinery on which their industries depended, he argued. "The gains that were made in Pittsburgh, Springfield, and Dayton," Mumford wrote "were realized largely in New York." The fact that ambitious young "provincials" rush to the metropolis in search of cultural vitality "comes to something like an attempt to get back from New York what had been previously filched from the industrial city."[23]

Mumford's call for deconcentrating the metropolis lay in a long tradition of antimonopoly thought alive to the dangers that corporate and financial consolidation posed to democratic life. He shared the populist conviction that democratic institutions cannot long endure without a more equitable distribution of the "common wealth." To that tradition, he contributed two original ideas, now generally overlooked: he argued for supplementing economic democracy with what he called "spatial democracy," and he broadened the term "wealth," to include both financial and cultural riches. Yet Mumford's legacy as an urbanist has been tied securely to debates about suburban deconcentration of the city, by Jacobs and her successors, who view him as an elitist mandarin, uncomfortable with the "organized complexity" and "street ballet" of dense urban neighborhoods. It is rarely acknowledged that Mumford, who is thus maligned for being unconcerned with class, applied his antimonopoly criticism to the metropolis itself, arguing that rising land values and expensive infrastructure led to higher taxes that together further impoverished the working poor or drove them out of the city at exhaustingly long distances from their work. Metropolitanism on such a massive and expanding scale, Mumford argued, took its highest toll on working people. It could not long endure economically or socially—nor should it.

Historians of urban planning look to Mumford's participation in the development of Radburn, a planned community in the New Jersey suburb of Fair Lawn, for clues to how his ideas applied in practice. In spite of Mumford's disappointment in the project, they find Jacobs's interpretation of his views borne out in suburban Radburn.[24] Neglected are the ways Mumford's idea of spatial democracy—his decentrism—could apply to the rest of the country, and particularly to smaller industrial cities.[25] Here, the idea of the ecological region is critical, for unlike "metropolitan systems of distribution," he argued, which are "anti-rural and therefore partial and lopsided . . . the regional system provides a place for every manner of human need."[26] Among the many problems with pushing growing metropolitan populations outward is that the suburb is "without the discipline of rural occupations," yet it also lacks the "cultural resources" of the "Central District."[27] The same was true of small cities flung willy-nilly across the Midwest where both agriculture and manufacturing had been industrialized: in a broad sense, these places too were "suburbs." Their inhabitants were "taught to despise their local history, to avoid their local language and their regional accents, in favor of the colorless language of metropolitan journalism: their local cooking reflects the gastronomic subterfuges of the suburban womans' magazines; their songs and dances, if they survive, are elbowed off the dance floor."[28] To flourish, he argued, cities of smaller size must stop copying abjectly the pseudo-national mass culture and draw instead from their own cultural traditions. And that can happen only if they remain decentralized in relation to one another, as well as in relation to large central cities, connected by transportation routes in a regional web respectful of environmental resources. Only by taking cultural, aesthetic, and technological matters in hand, in a spirit of democratic self-determination, could America's smaller cities conjure a true, diverse metropolitanism from the life-withering blur of machine-made uniformity serving the demands of the distant metropolis.

These were not the musings of a young man charting out ideas he would later feel obliged to retract. "The new form of the city," 71-year-old Mumford testified before Congress in 1967, "must be conceived on a regional scale: not subordinated to a single dominant center, but as a network of cities of different forms and sizes, set in the midst of publicly protected open spaces permanently dedicated to agriculture and recreation. In such a regional scheme, the metropolis would be only '*primus inter pares*,' the first among equals."[29]

Still, Mumford revised some of his original ideas. His hope for smaller regional cities, should they awaken to their natural and historical endowments, once lay in new technologies that had the capacity to decentralize the metropolis: automotive transportation and the electric grid. (Ever ecologically minded, he also argued that the electric grid had the potential to distribute renewable forms of energy, from hydropower to wind energy.[30]) Like virtually all of his contemporaries, Mumford assumed that the growth of smaller settlements would not come at the expense of the central city's population. By the late 1950s, however, he had come to see the egregious effects of the automobile on both city and countryside. Railing against highway planners who tore up the pedestrian urban fabric and diverted public transit funds to serve the car, he came down hard on their promotion of low-density, standardized sprawl that had nothing in common with the regional garden city ideal. "Our national flower," he observed acidly, "is the concrete cloverleaf."[31] By the late 1960s Mumford also tempered his enthusiasm for building new cities and recognized the importance of modifying existing settlements along garden city lines.[32] With these qualifications, the essential germ of Mumford's concept of the balanced small city in the ecological region—"in which agriculture, the extractive industries, manufacture and trade will be coordinated, in which the size of cities will be proportioned to open spaces and recreation areas and placed in sound working relation with the countryside itself"—remains even more useful today.[33]

Mumford often suggested optimal population figures derived from Howard, of 32,000 to 58,000, but he finally concluded that Howard's own numbers were "a shot in the dark."[34] What is most important, he wrote, "is to express size always as a function of the social relationships to be served," which requires attentiveness to urban scale, appropriate density, and environmental integrity. Without such limitations, cities lose their "social focus." Cities should not aspire to perfection or the simple "good fellowship" of small towns; they must be populous enough to facilitate economic creativity, class diversity, and the "drama" of the chance encounter. Small cities could be home to such vitality if they found a way to prevent "a few existing centers" from aggrandizing themselves on a monopolistic pattern" or, worse, aspiring to become one of them.[35]

If smaller cities were obscured by the way that Jacobs recast "decentrism" as suburban sprawl, they also fell victim to the sorry fate of regionalism itself, which Mumford spelled out in his 1967 congressional testimony. One of the movement's two best-known public projects, the Appalachian Trail, coordinated by Regional Planning Association of America member Benton MacKaye, had a decidedly conservationist, recreational intent that bore little relationship to urban planning. The other, the ambitious New Deal–era Tennessee Valley Authority, succeeded in converting water resources for energy use on a massive scale, but it failed to resettle the population displaced by the dams or to show much respect for the local culture. The National Resources Planning Board, established in 1933 to coordinate public relief, and therefore planning on all levels, created twelve regional planning boards but gave them no implementation authority.[36] Even worse, it made no effort to integrate regional planning with urban planning—"an absurdity," Mumford stated, reflecting a "compartmentalization" that continued for decades to come.[37] And instead of bypassing urban cores, "small or big," as MacKaye had called for, federal highway engineers enlarged on the excesses committed earlier by the railroads: "gouging through the center of the city and pre-empting its most valuable urban land for six-lane highways and parking lots and garages."[38]

Indeed, transportation planning—mainly to accommodate interstate highway building, which received still unrivaled levels of federal infrastructure support through the Federal Aid Highway Act of 1956 and its successors—became the default organizing principle for whatever remained of regionalism. Since 1962, federal transportation law has required "urbanized areas" of 50,000 or more to create metropolitan planning organizations (MPOs), through which state and local officials exercise discretion in spending federal highway dollars. In fact, the fundamental decisions have been made on the state level, with only informal influence provided by metropolitan regions—and the larger the metro, the greater the influence in statehouses. With the devolution of federal regulatory authority to state and local governments launched during the Reagan years, MPOs have acquired increasing responsibility for revenue streams and a variety of intermodal transportation forms. Studies show that although MPOs vary widely—in the ways they're constituted, in experience and technical proficiency, in relative staffing size—those that are large, serving regional populations of more than 1 million, and have performed well, "possess greater legitimacy as regional actors."[39]

And it appears that as regional transportation planners now prepare for a growing population over the next forty years, smaller cities could find themselves overwhelmed by the imperatives of their larger metropolitan counterparts—even as Lewis Mumford's ecological regionalism has found a growing number of advocates in the sustainability movement.

2

Megadreams and Small City Realities: Trafficking in Transportation Planning

"Washington sent money to Baghdad to build *its* electric grid," Hunter Morrison says to me emphatically across a conference table in his Youngstown State University (YSU) office. "That's money that could have gone to smaller industrial cities to restore *their* infrastructure."[1] I'm at the tail end of a weeklong fall 2009 trip to Ohio while recovering from swine flu, but Hunter holds my steady attention. I had already spent two days in Youngstown, learning about the status of the city's path-breaking comprehensive land use plan, Youngstown 2010, designed to reshape the shrinking city for a more productive, sustainable post-steel-industry future (discussed in chapter 4). I wanted to talk with Hunter, a principal architect of Youngstown 2010, about regional transportation development, particularly the much ballyhooed push for high-speed rail. I wondered, openly, whether high-speed rail—which advocates say is as necessary to future economic development as the postwar federal highway system was in the fifties—might further divert attention from smaller cities, which have much to offer a low-carbon economy.[2]

There's a bottled-lightning quality about Hunter Morrison. Grounded yet intense, with receding gray hair, a white mustache, and gold-frame glasses suggestive of his early-boomer demographic, he speaks in a low, confidential tone chopped by dry sarcasm and occasional bursts of ragged indignation. Chief planner for the City of Cleveland for twenty-two years—he stepped down in 2002 when his wife was elected mayor—Hunter now serves as YSU's director of planning and community partnerships. In that post he has, among many other things, led efforts to stitch the university's footprint into the downtown core as part of the city's larger economic development strategy. Looking beyond Youngstown itself, he also serves as an adviser to a raft of regional development organizations, including All Aboard Ohio, a passenger rail advocacy nonprofit, and America 2050, a nationwide foundation-supported initiative pressing for high-speed rail.

High-speed rail is intimately tied to the concept of the megaregion, itself a product of intensive globalization. The idea of cultivating the megaregion had origins in the late 1990s when, after twenty-five years of federal neglect, the Clinton administration began addressing what it referred to as a growing "crisis of the metropolis." Global competition, administration officials argued, wreaked havoc on urban centers that were already losing population, jobs, and tax base to exurban sprawl, a trend both economically and environmentally unsustainable. Advocates of what was then called metropolitan regionalism recognized that more than ever before, global markets functioned regardless of national, state, or municipal boundaries, giving rise to a new agglomerating economic geography. These ever-growing spatial units require "governance," they claimed, rather than government, consisting of stakeholders—elected and administrative officials, private investors, business leaders, advocacy experts, philanthropists, and community organizations—whose interests cut across political jurisdictions. Early on, governance debate focused on how to reintegrate cities more equitably into the networks of their larger and growing economic regions in the areas of transportation, infrastructure development, and job growth. In recent years, that debate has given rise to the concept of megaregions, enormous geographical swathes, or corridors, each grounded by at least two large metropolitan centers. The megaregion is fast becoming the new policy framework for sustainable planning, most notably in transportation.[3]

The most ambitious long-term proposal to emerge from the megaregional consensus is high-speed rail. Championed by America 2050 (among other coalitions), high-speed rail would form the circulatory transportation structure for an estimated population increase of 130 million over the next forty years. The idea here is to plan deliberately for concentrated population and economic growth in ten or so emerging megaregions, two of them in the Great Lakes and Northeast. Intended to repeat the successes of bullet train projects in Europe, Japan, and China, high-speed rail would connect big cities of 2 million or more that are anywhere from 100 to 500 miles apart, with the purpose of attracting commuters not currently well served at such distances by car, passenger rail, or air.[4]

America 2050's vision has many virtues, not least that it proposes a path toward sustainable practices for the 60 percent of the U.S. population already living within a megaregional scope—practices that include renewable energy development and upgrades to regional water management systems. High-speed rail is extremely expensive, however, not least because it would require new dedicated rail lines and hundreds of miles of tunnels. Its eleven proposed projects nationwide (the number varies in the broader literature) are estimated to cost anywhere from $500 billion to $1 trillion, but no one has yet calculated the full price in both dollars and disruption to the existing metro fabric and countryside. Building such massive infrastructure projects would create jobs in the short term. But in the end, it would provide commuter service only to megaregional knowledge workers and global elites who jump from metropolis to metropolis to ply their trades.

The case for facilitating megaregions through high-speed rail reveals just how little the proposal takes smaller cities into account. Transportation for America, a smart growth subsidiary affiliated with America 2050, for example, issued a platform for guiding the transportation reauthorization bill that called for "direct high-speed rail service linking our nation's largest cities" by 2030. "Smaller cities have needs too," the platform noted wanly, which it then lumped together with those of "towns and rural areas" and left the matter at that.[5] Elsewhere they aren't mentioned at all. It's hard to avoid the conclusion that unless they fall within a megaregional corridor, smaller cities are in danger of being cut adrift. Even those that are geographically fortunate in this respect, such as Janesville, run the risk of subordinating their bioregional strengths to the economic demands of megaregional development.

I asked Hunter Morrison, an enthusiastic proponent of both high-speed rail and Youngstown's future, if he shared any of my skepticism. Why not spend all that money on replacing, upgrading, and extending our passenger rail system in a true network instead of increasing rail speed in megaregions that already have decent rail, for privileged commuters? Where would that leave places like Youngstown? "I share your skepticism," he replied. "The idea of the megaregion is, rhetorically, something like the flavor of the day, like the 'creative class' was a few years ago. But it really doesn't have to be mutually exclusive with the idea of the metropolitan region on a smaller scale."

Hunter believes that high-speed rail is "an inevitability" thirty years out, its timing dependent on "demographics and the price of oil." But he also maintains that it's essential to develop commuter rail too, as well as light rail, streetcars, and bus rapid transit in cities with enough population density to support them. With these systems, he says, "frequency is more important than speed." Here, Hunter invokes postwar Britain's New Town planning initiative, based loosely on Ebenezer Howard's garden city idea. In contrast with the United States during the same period, the British built compact suburban communities without stripping out their public transit. "Our narrative is based more on national labor force mobility," he says, with little long-term public commitment to supporting specific urban places. "The British favored a more balanced approach to economic development and transportation systems, making it more difficult to abandon smaller places like, say, Cardiff [Wales], while also minimizing sprawl. Here the car became king."

Against big odds, Hunter has been pushing to bring high-speed rail to Youngstown. (The odds are even greater since Ohio elected a Republican governor opposed to passenger rail of any sort in 2010.) His case is based on reframing the old Steel Belt—Youngstown, Pittsburgh, and Cleveland—as a regional Tech Belt. "From this perspective, we are not a city of 85,000, but a bistate, three-city region of 7 million, with unitary labor markets and commuting sheds," Hunter explains to me by phone a year later. "If you frame your mental map this way, there are lots of jobs in total in the region, but it's hard to get to them by car." Current proposals call for running a high-speed rail line from Washington, D.C., to Pittsburgh, and another from Cleveland to Chicago. That means you couldn't travel directly from Washington to Chicago, from the East Coast to the Midwest, "which is ridiculous," he says. Hunter wants to see this gap filled by a high-speed rail corridor running from Pittsburgh to Cleveland through Youngstown, a historic "convening place" for area rivals, from steel magnates and union leaders to organized crime capos. Even if that doesn't come to pass, he argues, Youngstown should be reconnected to high-speed rail running between the two larger nodes through regular passenger rail, a service that hasn't been offered since 1977.

The Tech Belt idea rests on breaking down regional development into twenty megapolitan units (a term sometimes used interchangeably—and confusingly—with "megaregion").[6] This approach includes smaller regions such as the Carolina Piedmont anchored by Raleigh and Charlotte, or the "Ohio Valley," running from Columbus and Dayton to Cincinnati, as well as the bistate Tech Belt—scales that can learn more from one another than from a gargantuan megaregion model that excludes, marginalizes, or threatens to engulf the identities of many smaller cities, including Youngstown. To advance the Tech Belt's common interests, Hunter has been meeting with local chambers of commerce and community organizations, and bringing them together with their counterparts in Cleveland and Pittsburgh through the Regional Learning Network, established in 2008. "The concept of the region doesn't work at all well on the ground," he says, however useful it is for distant transportation planners and policy analysts. "If you approach the subject top down and tell people that the region is a place they have to belong to, they'll just ignore you." Instead, he observes, support for regional planning and cooperation "has to come from local leaders themselves through a process of discovering their joint interests in its internal logic."

Youngstown is doing everything in its power to keep body and soul together—pruning back its footprint, repurposing its brownfield sites, engaging in regional cooperation, and successfully incubating new businesses—but it still has the look and feel of disinvestment. "East Coast think tanks and academics tend to blame the cities themselves," Hunter tells me, "but much larger forces have been at work that *deliberately* gutted them, sending work overseas." He sounds weary. Having driven in from Cleveland 75 miles away, since there were no other viable transportation choices, he had been stuck in traffic at 7:30 a.m. The highway congestion in that corridor isn't likely to change anytime soon. It had become clear over the previous few months that leading proponents of high-speed rail, citing limited federal resources, were throwing all their weight behind making high-speed rail in the Northeast Corridor "a national priority." Since then, in late 2010, newly elected Republican governor John Kasich turned down $400 million in federal funding for high-speed rail in Ohio.

More modest ambitions for rail in the Buckeye State could prevail, however. In the 1980s, with the demise of Youngstown's steel industry, freight rail service was closed between Youngstown and the Port of Cleveland. Restoring one mile of track in the town of Ravenna would reconnect the two cities for both freight and passenger rail, infrastructure critical to the emerging Tech Belt region. It would cost $10 million—about the price of just one highway interchange.[7]

Road Warriors

As debates over rail and regionalism unfold, smaller industrial cities—both those that lie within emerging megaregions and those more far-flung—have reason to be wary of America 2050's vision. It has become a given that sustainability works on all scales. A major test of that conviction lies with transportation planning and land use policy, which lead to indelible alterations to the landscape and ecosystem. Now is the time to articulate the distinctive ways smaller industrial cities have been hurt by transportation planners in the past and to lay claim to a stronger role in such planning for a sustainable future.

Consider the disproportionately devastating effects of downtown highway systems in smaller cities—a subject that always puts me in mind of Hartford, Connecticut. Every time I drive through Hartford on the main route between Boston and New York, my shoulders tense up with anxiety—and anger. A small city of some 125,000 and Connecticut's state capital, Hartford hosts the intersection of two major highways, Interstates 91 and 84. That tectonic mash-up takes place right in the middle of downtown. The multilevel six- to-eight-lane tangle of concrete, asphalt, and steel stretches more than three miles. Navigating it is a sustained nightmare. Living anywhere near it—and who, in a city of this size, has any other choice?—must be a living hell.

In the years after the Civil War, Hartford became the wealthiest city in the United States thanks to the fortunes of the modern insurance industry. It was also among the most beautiful, with sturdy architecture, lovely urban parks, and great natural beauty springing from the banks of the Connecticut River. The city supported rich cultural institutions and was home to a vibrant publishing industry and literary culture that included Mark Twain and Harriet Beecher Stowe. Hartford was also a growing industrial city, manufacturing firearms, bicycles, automobile components, and textiles, and it attracted large military contracts during World War II.

Deindustrialization and suburbanization hit Hartford hard, repeating a pattern common in cities throughout the industrial Northeast and Midwest: the city itself soon became among the poorest in the country, with neighborhoods abandoned by the white middle classes filling with impoverished African Americans, and Puerto Ricans (and, as time went on, struggling immigrants from a variety of backgrounds). The process was accelerated by construction of those highways, completed in 1969. And again repeating a larger pattern of urban destruction, the interstates were built in the poorest of neighborhoods. Plans for creating a highway belt around the city were successfully resisted, particularly in the first-ring suburb of East Hartford, on the eminently reasonable grounds that the town already hosted too many highways. It was a rare instance of suburban participation in what came to be known as the freeway revolt.[8]

Early transportation leaders did not foresee the radical depopulation that urban highways would help bring about. Their rationale for running highways through urban centers—like a stake through the heart—was not to spur suburban development at cities' expense, although that was one of its many sour consequences. Their intention was no less nefarious, however. As articulated by the Urban Land Use Institute as early as the 1930s, the program was intended to equip the central business district for anticipated economic expansion while clearing the urban core of "blight." "Displaying a 'two-birds-with-one stone' mentality," writes historian Raymond Mohl, "cities and states sought to route interstate expressways through slum neighborhoods, using federal highway money to reclaim downtown urban real estate. Inner-city slums could be cleared, blacks removed to more distant second-ghetto areas, central business districts redeveloped, and transportation woes solved all at the same time—and mostly at federal expense."[9]

No one has yet undertaken a study of how—or whether—the freeway revolt played out in smaller cities. For that matter, I could find no secondary analysis of urban highway-building practices in cities of smaller size as a group. However, a 1961 consultant's report commissioned by the American Automobile Manufacturers Association, *Future Highways and Urban Growth,* provides a glimpse of how the highway lobby treated urban scale. It argued that four-lane highways (as opposed to those with six or eight lanes) were better suited to smaller cities not because of "urban form," which it dismissed as irrelevant, but because of lower population and therefore smaller traffic volume. But it also observed that "smaller cities have lower average land-use densities and relatively few destinations within walking distance" and thus greater car ownership in contrast with "large metropolises," where much daily business is transacted by foot. "It is, therefore, very desirable," the consultants advised, "to have freeway routes enter smaller cities rather than bypass them." Tragically, they viewed such an approach as preventing sprawl. "A bypass facility attracts new uses to itself," the study argued, "often decentralizing the community in a most undesirable fashion by encouraging new freeway-oriented shopping centers and industrial parks that may intercept movements to older centers."[10]

Many small-to-midsize industrial cities bent to the pressure of these arguments, along with the short-term promise of construction jobs, although each mile of highway occupied 24 acres of land, and each interchange, 80 acres—land that might otherwise have contributed to a city's tax structure. Many even welcomed downtown highways with enthusiasm: most Syracuse citizens (whose homes were not slated for urban renewal), for example, saw plans to drive Interstate 81 through downtown as a thrilling corrective to the decision, eight years before, to run the New York State Thruway north of the city rather than through it.[11]

It's all too easy to survey the wreckage: Hartford, New Haven, Springfield, Fall River, New Bedford, Worcester, Manchester, Concord, Albany, Syracuse, Rochester, Flint, Akron, Dayton—the list goes on. Buffalo paved over an Olmsted parkway to lay down Route 5. Just a decade ago, Peoria, which already had a bypass system, reversed course and drove an extension of Route 74 through the middle of the city. The rare exceptions—Muncie, Janesville, Rockford, and Youngstown come to mind—prove the rule. Their appearance, less forbidding in spite of the many misfortunes they share with their distressed counterparts, throws into high relief the disproportionate damage inflicted by urban freeways in cities of smaller scale. And as a result, today they are better equipped to rebuild their urban fabric.

If smaller scale was an important factor in the highway-driven destruction of these places, however, it is also a boon in the task of reconstruction: smaller cities stand to gain proportionately more by tearing down these monstrosities. These hulking feats of engineering hubris are now forty to fifty years old and facing the end of their life spans. Keeping them safe will require considerable expense. Of course, ripping out even small sections of these enormous rights-of-way (including their on- and off-ramps), replacing them with at-grade urban boulevards, and rerouting through traffic also costs a lot. But it usually costs less than rebuilding, and it costs much more to depress highways below grade, which may help quell excessive noise and light but has little effect on air pollution and, unless covered by an at-grade, landscape-level "lid," does nothing to restore the integrity of the urban landscape.[12]

John O. Norquist, president of the Congress for the New Urbanism (CNU), has made urban highway teardowns one of his chief missions through CNU's Highways to Boulevards initiative. As mayor of Milwaukee from 1988 to 2004, he launched a project to tear down a mile-long elevated highway spur, the Park East Expressway, that ran through the heart of the city. Norquist estimates that it cost about $55 million less to dismantle the thing than to fix it. Completed in 2009, the teardown opened up 60 acres of the city, once shadowy and ugly, for infill development. Surrounding neighborhoods formerly cut off from one another are slowly reintegrating around what is now an at-grade boulevard, which is much less expensive to maintain than the old highway. Norquist's bet is that this gambit will pay for itself over the long run by increasing the tax base.[13]

Norquist, who likens Hartford's highway complex to "a turd in the punchbowl," has been on a crusade to promote highway teardown projects in cities both large and small. Such projects face enormous odds, not only fiscally but politically: governors have control over the distribution of federal transportation funds, and they are required to allocate most of that money to highway development, in what Norquist calls "sprawl by law." To those who argue that highway expansion follows the will of the free market, he argues that the power of the highway lobby creates the market in what he calls "communism for highways." Asked whether poor minority neighborhoods resist highway teardowns, fearing a reprisal of the urban renewal programs that accompanied their construction in the first place, he replied that the highway lobby plays the race card, but most people don't buy it. Plans to tear down 2.2 miles of Interstate 10 through New Orleans, he noted, was greeted by the nearby black neighborhood with universal acclaim. More common, he says, is resistance to such efforts as street narrowing and neighborhood detailing, which minority residents view as the advance guard for gentrification—with much justification.

Although Norquist does not stress the proportional advantages of highway teardowns for smaller cities, some are taking to heart his leadership in what might be called the freeway revolt, version 2.0. Trenton, New Jersey, was far enough along in the process to apply for 2008 federal stimulus funds (the funds were denied). Others are just beginning the process. In 2008, the Onondaga Citizens League published *Rethinking I-81*, a report on the feasibility of tearing down a particularly hideous portion of elevated freeway in downtown Syracuse. Robert Simpson, head of the area Metropolitan Development Association, told me that it will likely happen though probably not completed before 2030, due to the long six-year intervals in federal highway authorizations. And, as usual with transportation planning, much depends on future gubernatorial leadership. For still other cities, these divisive eyesores pose a host of engineering problems. The sustainability-minded mayor of Akron, Don Plusquellic, would love to rid his city of the alienating concrete blight that runs through the middle of downtown, but it is set so far below grade that one city planner half-joked to me that they had considered filling it with water and turning it into a pond. Nevertheless, one spur could come down, but the Ohio Department of Transportation is completely unsupportive (typical of state departments of transportation). From an engineering perspective New Haven has a chance to displace Interstate 95, but Hartford's prospects are grim, as are those of Springfield, Massachusetts.[14]

In spite of the many obstacles, small industrial cities' urban cores arguably have even more to gain from dismantling these misguided remnants of an earlier era than big cities do. It's usually not only cheaper than repairs, but at-grade motorways are much less costly to maintain than elevated freeways, with their numerous bridges and ramps.

Moving the People

Smaller scale is a strength and an advantage in transportation planning in other ways too. If smaller cities don't have the density or the funding to support light rail currently, they have the flexibility to develop bus transit in imaginative new ways. In fall 2008, as the United States was on the verge of economic collapse and cities were in fiscal free fall, Rochester, New York, did the unthinkable: its bus system reduced rider fares from $1.25 to $1.00. The Rochester Regional Transit Service took the measure at a time when fares for mass transit were going up across the country, in large part to pay for debt service. At the same time, New York's Metropolitan Transportation Authority was in debt to the tune of $24 billion—among the highest of any other public entity in the entire nation.

How did Rochester pull it off? First, the regional transportation authority, which had been carrying a substantial deficit, received a hefty increase in state aid and banked its surplus. Then it negotiated a range of subsidy agreements with local businesses and educational institutions, most critically with the Rochester City School District. The authority cut a deal to provide transport to almost all public high school students at a cost of $2.22 per ride. That might sound exorbitant, but had the school district contracted with a private bus company, the annual cost would have been $2 million, or 20 percent higher. The authority also canceled routes with low ridership but kept some open on reduced schedules in exchange for subsidies paid by institutions and apartment complexes that wanted bus service.[15]

A year later, in fall 2009, bus ridership was up by 2 million—a twenty-year high during a time when the city's population had dropped 15 percent—and the city was able to purchase fifty new eco-friendly buses. The new buses run on clean diesel filtration technology, which is expected to yield an annual $15,000 in savings in fuel and parts per bus, for a total savings of $750,000 a year. Meanwhile, the fare has been kept at $1.00, and the system carries no debt. Plans were also in the works to outfit the fleet with Wi-Fi and to provide city buses with signal light priority.[16]

One could argue that comparing Rochester's 2008 $62 million public transit system with New York's $6 billion budget, which covers both its bus and subway systems, is like comparing apples and oranges. But that is precisely the point. The automaker, oil, and tire industries' coordinated effort to systematically uproot the nation's trolley and interurban system, beginning in the 1930s, took its greatest toll on small-to-midsize cities. Nevertheless, it left the cities that were most harshly affected less burdened with the capital expense of maintaining costly light rail today. As a result, local transportation planners in smaller cities have more flexibility to find markets for public transit users and to shape their routes accordingly. Moreover, in a city the size of Rochester (with a population of about 207,000), it's simply easier to do business with institutional power brokers, both public and private. Navigating the thicket of overlapping interests, institutions, and constituencies in, say, New York or Chicago, makes it difficult to negotiate the kinds of arrangements made by the Rochester transportation authority.

None of this is to say that smaller cities should give up on restoring light rail—hardly. But as with all big transportation projects, it will have to await a serious federal policy commitment. Cities cannot act alone with projects of such magnitude. Support for light rail is suspended for now in favor of funding highway expansion and research in clean-fuel and battery technology in hopes of keeping automotive travel viable in a low-carbon future. In anticipation of future population growth, smaller cities have good reason to step up lobbying for light rail. In the meantime, however, they have entrepreneurial advantages in transportation planning that large cities lack. Rochester has demonstrated one way to do it. Now, if only the Flower City would tear down the unsightly, pedestrian-hostile Inner Loop encircling its downtown core.

City or Suburb?

It's tempting, of course, to do nothing to substantially improve mass transit and simply wait until national politics and the market shift course. For smaller cities that lie within conceivable reach of a huge metro area, there's yet another temptation: to build more highways in anticipation of integrating more fully with megaregional growth—that is, to build outward rather than upward.

Rockford, Illinois, which is about 65 miles northwest of Chicago's O'Hare International Airport, is one such city, and it has been going through an identity crisis in recent years. "Do you think Rockford is a Suburb or a City?" runs one discussion thread on City-Data.com, a puzzled sentiment echoed by others during my visit there in November 2009. For those concerned with civic identity and urban planning, a great deal depends on how one answers that question. For although Rockford itself has been losing population, the adjacent area has been growing steadily, by some 35 percent, over the past twenty years, and it is in danger of losing its identity to the Chicago metropolitan area, dubbed Chicagoland.

A medium-size city of some 155,000, Rockford has fallen on hard times. Once a stately little burg, it was the girlhood home of Jane Addams, who graduated from the prestigious Rockford Female Seminary in 1881. At that time, the city was a center of furniture making, and until recently, it had an unusually high number of family-owned manufacturing businesses specializing in machine tools, auto components, fasteners, and toys. Since the mid-1990s, however, several list makers have called it one of the country's worst cities to live in. In a strange development pattern, the predominantly white east side stretches for seven miles along East State Street until it meets with a length of I-90 called (without a trace of irony) the Jane Addams Memorial Tollway, where it intersects with I-39. The city began annexing this land, building out from the center, soon after I-90/39 was completed in the 1960s, in fear that it would lose retail business to an early shopping mall built on the other side of the highway in the affluent suburb of Cherry Valley. As a result, the East State Street corridor resembles typical suburban sprawl even though it lies within city limits. Until recently, state law forbade the inclusion of sidewalks along roads upgraded with federal dollars, so there are no sidewalks along East State Street—a state of affairs the city is now beginning to redress.

By far, the bulk of the black and Hispanic poor live on the west side of the city, which reflects decades of neglect, with abandoned commercial and industrial buildings, dilapidated housing projects, and a huge inventory of vacant properties. The city has devised an ambitious West State Street corridor plan to remedy that too. Years ago, the Rockford metro area was one of few to build a highway bypass around the city. The new plan seeks to develop the four-mile blighted area of West State Street that runs between downtown and the Route 20 bypass (which lies beyond the city's limits), with the idea of providing more retail services for what it refers to as its low- to moderate-income "underserved market." When I visited Rockford, West State was a gritty four-lane stretch of road lined with houses, storefront churches, and commercial structures in a variety of conditions. The plan calls for tearing all that down, removing on-street parking, widening the road, adding a meridian, and anchoring each end (including the greenfields sitting on the county side) with commercial development where the city hopes to attract a big supermarket and several drugstores. Between these nodes, the city says it plans to create parkland along both sides of the corridor, consistent with its already bountiful system of parks. Its plan, however, foresees the possibility of filling in the widened road with more retail, including fast food outlets. It also lays out in intricate detail the increased automobile traffic the widened road will accommodate. All things considered, the plan looks more likely to result in sprawl rather than parkland—which the city doesn't seem to need in any case.[17]

It's not clear how fully the West State Street corridor plan will be realized. The original study on which it is based was completed in 2002, and the city is in even more dire fiscal straits now than it was at that time. What is clear is that the $4.25 million Illinois Department of Transportation grant that prompted the city's development study is going into effect. In November 2009, the city (with the county) was about to proceed with the first stage of the road project: to clear the existing buildings and widen West State Street. Plans are also in the works to offer highway expansion and commuter rail between Rockford and Chicago's northeastern suburbs.[18]

Rockford seems to be hedging its bets. Is it a city, or is it a suburb on its way to merging with Chicagoland? Either way, should it replicate the suburban design standards of the late twentieth century? Or should it push for something more appropriate to the low-carbon twenty-first century: compact, transit-oriented, pedestrian-friendly urbanism that discourages further development in the adjacent countryside, which is blessed with some of the most fertile soil on earth? Rockford, like all other cities—and especially smaller ones—is at the mercy of multiple county, state, and federal initiatives, not to mention the caprices of global capital, and it has only so much control over its destiny. But it does have some. By all accounts, the city's ambitious young mayor, thirty-five-year-old Lawrence J. Morrissey, elected in 2005 as an independent, has some good ideas for developing green jobs and rebuilding an ecologically sound infrastructure. It would be a shame of incalculable consequence to sacrifice those gains, in both environmental and aesthetic integrity, to megaregional sprawl.

3

"It Takes the Whole Region to Make the City": Agriculture on the Urban Fringe and Beyond

We are a long way from Ohio, Illinois, and Wisconsin now. It's April 2009, and the Harvard University Graduate School of Design is hosting a three-day conference, Ecological Urbanism: Alternative and Sustainable Cities of the Future. This interdisciplinary gathering of faculty, students, and urban practitioners from across the globe marked a turning point for the school. Until now, its identity had been inseparable from the renowned Spanish architect José Luis Sert, who served as dean from 1953 to 1969. Sert, who was mentored by Le Corbusier and later served as president of the Congrès International d'Architecture Moderne, was a European modernist of the highest order who called for integrating architecture with urban planning in what he was the first to call "urban design." In practice that meant putting lots of concrete in the service of streamlined superblock buildings with unadorned facades and sweeping plazas from which to behold them. Bold urban form had remained a preoccupation at Harvard, more recently in brash works of ironic postmodern eclecticism. "Ecology," for many of these folks, had a whiff of antimodernism about it, of formless authenticity reminiscent of hippies and back-to-the-landers.[1]

Not any more: Harvard was catching up and had coined the term *ecological urbanism* to signal its intent. As conveyed by the conference presentations, ecological urbanism encompasses carbon-neutral building design, green infrastructure engineering, and urban farming. At its most utopian, the conference covered from-scratch eco-cities, such as MADSAR in Abu Dhabi, and vertical farming incorporated into tall buildings; at its most prosaic, it explored practices such as how to manage stormwater runoff through green roofs and swale-based drainage and looked to the ingenuity of the world's poor for design ideas.

In spite of the jargon of postmodernism and high-tech engineering, there was much of interest here, especially among some of the landscape architects working on a regional level. But the general tenor of the event seemed both techno-futuristic and excruciatingly green—the latter, perhaps, in compensation for Harvard's past modernist sins. It wasn't clear how much these innovations might help smaller industrial cities get out from under decades of environmental degradation and economic neglect. At the time, I was at an early stage of the project that resulted in this book. I had come to the conference with a vague sense that the salvation of smaller cities lay in making sustainable use of their rich assets in agricultural land and water, intact urban infrastructure and historic building stock, and now-shuttered manufacturing facilities, that their smaller size—as cities—was itself an asset. "Ecological urbanism" didn't speak much to any of these considerations.

So I brightened when Andrés Duany, a cofounder of the new urbanism, made a surprise appearance. He was there to push for what new urbanists call the rural-to-urban transect, a design concept that divides a metropolitan region into six transect grades—from forestland, through rural and suburban settlements, to urban residential settings, to the dense urban core—and then proposes a loose hierarchy of densities and building forms appropriate to each of them. Duany's transect theory struck me as particularly suited to smaller industrial metro areas. Of Cuban descent, with meticulously cropped gray hair and a trim build clothed in a tan suit of exquisite tailoring, Duany cut a mercurial figure and seemed just a little out of place in this gathering. His droll yet straightforward speaking style called to mind the thrust and parry of swordplay: where the academics spoke of their "interrogations" and "provocations" (translation: "questions"), Duany *was* provocative. Well known as a brilliant promoter of the new urbanism long before the formation of the Congress for the New Urbanism in the early 1990s, he had clearly come to the conference to make a theatrical splash. Later, when I asked him why he was not on the published program, he replied that he would not have accepted such an invitation: "I do not like to be one among many."

I couldn't help but smile. Duany may have been engaging in a bit of self-ironic posturing, but he was very much in earnest about the transect. As was the case with Lewis Mumford and his circle of regionalists in the 1920s and 1930s, new urbanist transect theory is indebted to Scottish biologist Patrick Geddes's idea of the ecological region. Geddes refined older ways of classifying and visually mapping out, through cross-section transect diagrams, the relationships between plants, animals, and minerals and each of their natural environments. In two path-breaking moves that now seem commonplace, he stressed that habitats are symbiotic—that each grade of the transect, while discrete, needs the others for a region to thrive as an ecological whole. Geddes then added a social dimension to each transect grade, aligning occupational work with its associated natural resources, such as mining in the mountains or fishing near the water's edge). "In short," he wrote, "it takes the whole region to make the city."[2]

Oddly, though, Geddes's socialized transect only implied the city and did not examine good urban form in relation to the environment. The new urbanism fills in both omissions: it brings transect analysis all the way into the city, and it offers design strategies for each of the six ecological transect grades comprising the region. To that end, the new urbanists seek to replace the drab uniformity in commercial and residential building standards found in both the city and the suburbs with an infinite variety of designs suited to the idiosyncrasies of local conditions. Each transect grade therefore requires customized zoning and building codes—or form-based codes—devised to reflect the architectural, economic, and natural histories of particular places consistent with good urbanism.[3]

Good urbanism, by Duany's lights, requires compact, walkable streetscapes and convenient public transit, with provisions for mixed-use buildings constructed on "traditional" scales for multiple activities. Its modest aesthetics is historically grounded in the City Beautiful movement and the urban designs of John Nolen and Raymond Unwin. And it was everywhere destroyed by the introduction of single-use Euclid zoning (so-named for the 1926 Supreme Court case, brought by the city of Euclid, Ohio, validating the practice), which chopped up the American landscape with relentless vigor during the massive housing expansion after World War II. The purpose of Euclid zoning, a response to the pollution and congestion of the industrial city, was to separate residential areas from those reserved for production, both industrial and agricultural. It resulted in what Duany calls "modernist transportation and zoning": arterial highway systems connecting now-separated land uses; sidewalk-less, auto-dependent suburban tract, office park, and mall development; and sprawl into rural and forestland areas. These "standardized products" are not the consequence of neutral free-market choices, he argues, but of the sixty-year triumph of modernism, whose principles are routinized in our lending institutions and realtor tastes, our transportation engineering schools, and even our environmental zoning, which too often restricts sustainable urban planning. Form-based coding, of which the new urbanist transect-based SmartCode is one version, is intended as an alternative to the twentieth-century practice of rigidly separating land uses—to the detriment of the public realm as reflected in built form.[4]

"Fortunately," Duany announced to the Harvard gathering, "I am here to restore the standing of plain old good urbanism." He began by chastising attendees for paying attention only to big-ticket clients (mainly in Chinese and Middle Eastern megacities) or to globalization's "victims," whose shantytowns they praised as models of urban social and biodiversity. Missing altogether—"for ideological reasons," Duany claimed—were American middle-class "customers" and the developers who seek their business. This neglect is crucial, he argued, because the American middle-class's high-carbon way of life has been largely responsible for our environmental predicament, and its consumer expectations are now being exported across the globe. This is a critical moment, Duany insisted, targeting his comments at the increasingly influential self-identified landscape urbanists in the crowd who, he argued, seek to restore natural ecosystems at the expense of well-designed buildings and neighborhoods fit for human thriving. Inundated with environmentalist advocacy for decades, the American middle class had slowly become ecologically minded and had reached a tipping point, Duany observed. The question now is how to blend this newly "green" middle-class consciousness with the creation of good urban form. The biggest challenge to those who care about a low-carbon future, Duany observed, lies with the peculiar nature of American environmentalism itself, which idealizes untrammeled wilderness, places human beings "outside of nature," and seeks to "green" everything indiscriminately. As a result, Americans are inclined to bring nature into the city, or to "ruralize" it, and to "greenly aestheticize" suburban form. This is all wrong, Duany contended, a series of "transect violations." Until the urge to green takes into account variations of urbanism, Americans will continue down the path of sprawl—with its disastrous environmental and socioeconomic consequences.[5]

This chapter and the next one are devoted to the place of agriculture and food production along the transect. Long criticized for building planned communities on open land, or "greenfields," and thus contributing to sprawl, the new urbanism has only gradually taken urban infill development and agriculture seriously. Yet it does so convincingly and in ways that could be particularly useful, it seems to me, to smaller industrial cities (even though the high volume of vacant properties in many of them would require commensurate tailoring of form-based codes).[6] This chapter discusses agriculture in relation to the rural and suburban transect zones. In chapter 4, we'll move on to consider farming in the urban core, along with opportunities to restore urban ecosystems through green infrastructure. Starting from the outside-in makes sense from an ecological systems standpoint. Although the novelty of urban farming has garnered a great deal of recent attention, true regionalism and the transformation of agriculture must begin with sound rural practices.

Smaller cities hold a number of advantages in developing what new urbanists are now calling agricultural urbanism in their rural and suburban areas. In a 2006 article for *Places* magazine, "Building Community across the Rural-to-Urban Transect," Charles C. Bohl and Elizabeth Plater-Zyberk (Duany's wife and new urbanism cofounder) laid out the principles of transect zoning and considered objections to it. The most compelling criticism, and one the authors couldn't fully counter, came from Kansas City architect and planner Kevin Klinkenberg: "The problem I'm having is: these tools, while perhaps getting us 80 percent of the way there, are not yet sophisticated enough to deal with the planning issues of large, contiguous areas of urbanism, especially in our older cities. . . . The Transect and SmartCode [customized zoning for appropriate urban design form and mixed uses] are exceptional for dealing with small towns, TND's [traditional neighborhood developments], and smaller cities. But for the cores of our larger metropolitan areas, we're not really there yet."[7]

In a February 2010 telephone interview, I asked Kevin to elaborate on these observations with regard to smaller cities in the Midwest, where most of his projects are located. In good new urbanist fashion, all of his commissions require an intensive community visioning process, or a series of charrettes, that works "best in smaller jurisdictions," he said. "There are so many issues involved with changing mind-sets, developing new regulations and codes, and twisting arms. It's easier to show people what different neighborhoods along the transect can be like. In big cities, things are more complex. There's a blurring of lines between neighborhoods and within corridors. Plus, the market is so big that things can change rapidly." He pointed out that Plater-Zyberk also found this to be true while working recently on a SmartCode project in Miami.

By contrast, Kevin observed, smaller cities have "different employment and growth pressures," and so "development and redevelopment happen at a much slower pace," making it easier to gain consensus and to see projects through. Cedar Rapids, Iowa, with a population of 128,000, for example, hasn't changed much in thirty years, he noted. "The year-to-year increment—even decade-to-decade—is much slower." As far as agriculture is concerned, he says, "these cities are sprawling at a slower rate too," which drives down the cost of farmland. That opens up all kinds of opportunities for rural and suburban agriculture at a time when interest in local food is growing in cities both large and small.[8]

Ruralizing the Countryside

One hazy temperate day in November 2009, I took a 20-mile drive east out of Peoria to the tiny hamlet of Congerville, Illinois, to visit Henry Brockman, who had been making a living as an organic farmer for seventeen years. I was in the rural transect zone now, driving down recently upgraded I-74. It was the end of the fall harvest season, and the air was sweetened by the scent of ethanol, at times cloyingly so. I'm reminded of the Wonder Bread factory I toured with my fifth-grade class. Once I got past an enormous Caterpillar distribution center in Morton, all was corn and soybeans—but mainly corn—anchored to the low rolling landscape by an occasional grain processing facility. This is what monoculture looks like. I'm from a dairy and corn-growing region in upstate New York, but I'd never seen anything quite like this before, not on this scale.

Monoculture. Michael Pollan had popularized the term in *The Omnivore's Dilemma*, published in 2006, unleashing what fast came to be known as the local food movement. You didn't have to be a particularly discriminating eater to become truly alarmed by what Pollan had to say. He showed in great detail how the system of commodity agriculture—corn and soybeans—had taken over our food system with mounting taxpayer subsidies since the 1970s. How the arrangement keeps prices low to the supreme advantage of gigantic industrial food monopolies—ADM, Cargill, and Tyson, for example—that force ever-larger yields and dictate how large livestock-raising factory farms conduct their operations. How corn finds its way into everything, from less nutritious, high-corn-fructose-addled food products to the diets of grass-eating cattle that now require antibiotics to stay healthy enough for slaughter. How seed monopolies such as Monsanto and DuPont cornered the market on genetically engineered hybrids in a coercive system that makes it nearly impossible for farmers of any size to avoid their expensive products. And then there were the huge federal corn ethanol subsidies that began in the 1980s. But for those unmoved by concerns about nutrition, or even monopoly control or animal cruelty, another crucial part of Pollan's argument rang a bell: the entire system is drenched in oil, from petroleum-based fertilizers and pesticides, to long-distance transportation costs, to machine-based field control. And it is unsustainable.[9]

Most of these arguments had been around for a long time, gaining critical mass in the 1970s with the organic farming movement. They found their most eloquent moral and spiritual expression in the work of philosopher-farmer Wendell Berry, most comprehensively in his 1977 book, *The Unsettling of America: Culture and Agriculture*. In an era of climate change and open warfare in oil-producing nations, however, Pollan's book struck a nerve—or, rather, more nerves.[10]

The local, sustainable food movement is now bigger than ever before. In large cities such as New York and Chicago, the locavore market is huge and has received much publicity. In New York City, the number of licensed farmers' markets grew from 32 in 1995 to 109 in 2009. The market for fresh locally grown food has also grown in smaller cities, towns, and suburbs, though the national media rarely report on it. Yet taking into account the high price of land and population density in big metropolitan areas, the metropolis might not be well suited to local, sustainable agriculture on anything more than a broad regional scale. A 2010 study commissioned by the Manhattan borough president Scott Springer proposes, sensibly, that New York source a substantial portion of its food 200 miles out, in upstate and Long Island farms, rather than importing it from across the world. Opening New York's 8-million-mouth market to the state's farmers would keep agricultural profits circulating within the state's economy, the study argues. It would also benefit the upstate cities that anchor farm economies, which themselves provide a market for truly local food. Huge obstacles lie in the proposal's path: the food industry and global shipping interests that profit handsomely from the status quo. Whether this particular proposal pans out, however, it suggests that the political will of the local food movement is reaching policymakers and could lead to a brighter economic future for smaller urban areas situated in the fertile soil of the Northeast and Midwest. It seemed to me that Henry Brockman, whose farm is located 20 miles equidistant between Peoria and Bloomington-Normal, might be able to shed some light on this.[11]

Henry is a diminutive guy in his early forties, whose weary blue eyes easily crinkle into a smile. If you knew of him only online, through his Web site Henry's Farm—a marketing tool, after all—you'd miss his most striking trait: his modesty. After introducing me to his wife, Hiroko, we ate spice cake in his open kitchen overlooking the Mackinaw Valley. A small-city urbanite, I was a little new to the details of agriculture, and Henry patiently answered my many questions. He started his farm in 1993, after going off to college and wandering around for a few years, on 3 acres of land. His parents gave him the land in an intergenerational land transfer typical among farmers. (Atypically his father was a genetics professor and his mother a nurse; they ran a hobby farm on their 50 acres of land.) Since then, the tillable portion of Henry's Farm has grown to 24 acres, half of it in vegetable production each year: this is a small farm. Henry originally intended to sell in Peoria and Bloomington, but there was no market—no farmers' markets, certainly, and "no taste for anything but potatoes, sweet corn, and tomatoes. *Why you bringing all these greens here?*" So early on, he began schlepping his produce up to a farmers' market in the northern Chicago suburb of Evanston, 150 miles away. "Chicago was supposed to be a stopgap," he said, before developing his business in the smaller cities closer by. But he made a profit in his first year, and he has been doing it ever since: twenty-six Saturdays a year, he gets up at 1:00 a.m., drives three hours north, sells to and visits with his customers all day, then drives home. A few years later, he developed a community-supported agriculture (CSA) arrangement in Bloomington-Normal that has grown in size. But he estimates that even today, two-thirds of his income comes from Evanston. In 2009 Henry's Farm earned more than $200,000 in gross sales.[12]

As anyone with even a rudimentary knowledge of agricultural economics knows, finding and maintaining a market is crucial to successful farming—as with any other business. Henry's market consists of people who care about sustainably cultivated, nutritious food. In Evanston, where he vends his vegetables under six tents, he sells everything he grows and his success is legendary. Each year at the beginning of the season "without fail," he tells me, a neighboring seller asks him if he's going to buy more land, expand his operation, and sell at six markets a week. "Why would I want to do that?" he asks. If his business grew any bigger, he wouldn't have enough time personally to grow and to sell and would be forced to hire out the work. As it is, his operation is truly a small family farm. His parents, wife, five siblings, and three children pitch in. He also has two apprentices and a part-time farmhand. He easily makes enough money.

Henry argues that truly biodiverse, sustainable farming must be undertaken on a small scale in close proximity to consumers. It requires careful cultivation of natural nitrates in the soil through crop rotation and regular tilling. He does this by alternating vegetable crops with clover or alfalfa legume cover crops, rich in bacteria as they decompose, over two-year intervals. It's labor-intensive work, but the result is soil teeming with microbes and nematodes and a form of nitrogen easily pulled up by plants rather than leeching into the groundwater. "My soil exudes the rich, loamy smells of yeast and life," he has written. "It crumbles softly in your hand, held together in clumps by the sticky glue of organic matter. My neighbor's soil has been sterilized by chemicals. It smells of nothing. It shears apart into hard-edged chunks and finally into fine dust."[13] Most of the big organic growers certified by the U.S. Department of Agriculture (USDA), whose produce is sold in supermarkets, don't treat the soil much better. Although they avoid synthetic fertilizers and chemical pesticides, they engage in monoculture, using high-nitrogen fertilizer and excessive irrigation that depletes the soil of its organic compounds. They also engage in a market system that puts an average of 1,500 miles between farmer and consumer.

Anyone who has read Michael Pollan and other critics of the industrial food system is familiar with these arguments.[14] What I wanted to know was whether the recent explosion of the local food movement might induce him to revisit his original plan to market to the smaller cities of Peoria and Bloomington, which were much closer. Had the world he had come to know so well after seventeen years of experience been Pollanized? "Well, it's mainly affected the consumer side, and maybe the new farmer side as well," he replied. His wife and parents were pressuring him to take advantage of the growing demand for locally grown organic produce and to confine his sales to Bloomington, where he could now easily pick up more CSA buyers. He had done the math and figured he could maintain his good income if he expanded from 200 to 800 buyers. But he decided to wait on that and considers it "a retirement plan." "I love going to Evanston," he said. "I've known some of these people—the foodies—for years, and they're interested in trying new things. Then there are the new immigrants interested in finding their native produce, like the short, round Russian ladies pointing in disbelief at the sorrel, or the Japanese overjoyed to find gobo [burdock root], or the Chinese thrilled by the winter melon." Besides, with CSAs you have to streamline what you sell. "They're less interested in trying things like Egyptian spinach and salsify root. This is central Illinois. They grew up on potatoes and sweet corn."

And that brings us back to Peoria. What about selling there? The market for local food had grown in Peoria too, but Henry was reluctant to plunge into it anytime soon. Here I learned yet more about agricultural economics. There are now three farmers' markets in Peoria, including a brand-new one on the developing riverfront. The oldest and largest, Metro Centre, is open six days a week. It, however, allows reselling—where a stand proprietor buys wholesale, often from a distribution center or from groups of distant farmers—which drives down the price. Local farmers can't compete, and buyers usually end up getting the same produce they could find at the supermarket, often at a higher price. Although many farmers' markets have grower-only rules written into their bylaws, they are regularly skirted. He also finds the practice in Evanston. West of Chicago in places like Rockford and Janesville, the price of land has soared due to exurban development pressure, and so has the agricultural tax rate. So it's economically difficult to maintain a small, sustainable farm even 60 miles out. In fact, a lot of the sellers come from the rich farming area of southwestern Michigan and make the long 100-mile-plus trip around the bottom of Lake Michigan to get their goods to the Evanston market. In one long trip, sellers often haul produce from a number of different farms. It's all too tempting in both cases, says Henry, to ignore the grower-only rule. He's able to make it work because he has a loyal base of customers. He knows many others who have "gotten out," including his Congerville neighbors at the Wettstein organic meat farm who tried selling at the Peoria Riverfront market. They couldn't make the numbers work.

The lesson here seems to be this: local farmers would like to serve the markets clamoring for local food in smaller cities like Peoria, and they can afford to do so given their relatively low land values and real estate taxes. But to entice them, consumers have to pay closer attention to how their markets are administered. "Farmers will do what makes them money," Henry tells me. "They follow demand and learn mainly by talking with each other. Word is getting around that sustainable agriculture is profitable and that you don't need much land to do it." The USDA extension service has been gradually moving in this direction too, so the support is there for those who want to make the shift away from ethanol. Over the past five years, for example, the University of Illinois extension school has been testing plots in organic production. "It's pretty funny to see the *exact same people* who used to come around pushing for herbicides and traditional fertilizers now talking about organic alternatives," he says. "A lot of it is just rhetoric. They mainly just want people to stay in farming, and they're looking only at the economics. But that's okay."

The older, entrenched farmers who on average have a thousand acres in grain production are pretty slow to move, however. On the production side, the local food movement has had the biggest effect on new farmers. It doesn't cost that much to get started: good farmland goes for about $5,000 an acre in central Illinois—it's more expensive in Ohio, cheaper in southern Illinois—and when it is properly cultivated, farmers don't need much to grow a lot of produce. One of Henry's apprentices, Matthew, is trying it out. His father owns 500 acres devoted to corn, soy, and cattle; Matthew is putting 5 of those acres into organic production. If he does it right in terms of both marketing and soil cultivation, Henry estimates that Matthew should double what his father would make on that land within the first year and ten times as much not long after. Another apprentice who hails from a suburban Navy family that traveled the world and owned no land just started a farm farther south, outside Carbondale, where the land is cheaper.

It is all too easy for globalization's cheerleaders to regard small operations like Henry's as microfarming, with no chance of saving old Rust Belt cities and their rural hinterlands in the face of globalization. These small enterprises are just drops in a vast ocean of corporate commodity agriculture where farms must grow bigger and ever more efficient to compete with farmers working cheaper land elsewhere in the world. This argument assumes, however, that oil and water are inexhaustible resources and that intercontinental shipping costs will remain low. For now, Henry's apprentices are seeking a good living while hedging their bets that local markets for fresh food will continue to grow. "Commodity farming has been around since the 1850s," Henry observes. "It's going to take many generations to restore the soil and skills we lost." To that end, in 2001 Henry's sister Terra launched a nonprofit, the Land Connection, which helps heirs to farmland put agricultural easements on their property and offers incubator training programs for farmers who want to replace monoculture with sustainable, organic practices. The Land Connection is more successful with younger farmers than with farmers sixty years or older, who currently work on an industrial model, she told me in a telephone interview. Some of these newer farmers (who are either new to farming or coming back to family businesses they had left behind) "are making the decision to sell in smaller cities—not just in Chicago—where the demand didn't exist fifteen years ago." What is needed to meet and expand that demand, says Terra, "is really quite simple: land, trained farmers, local processing facilities, [which disappeared in the sixties], and logistical transportation."[15]

Before I left, Henry disappeared for a moment and came back with a slip of paper. "I think it's important to address the feed-the-world argument," he told me. By this he means the idea that industrial agriculture, with its genetically modified food and petroleum-heavy inputs, is necessary to keep pace with population growth, especially for the world's poor. "Yeah, so I ran some numbers." Henry figures that he feeds 250 people per acre for six months of the year (his marketing season), or 125 people per acre per year. At his scale of approximately 10 acres, that means that it would take 100 small farmers to feed a city the size of Peoria or Bloomington, each with populations of around 125,000. He then extends the analysis to show that two-fifths of 1 percent of Illinois's farmland (currently 27 million acres) could feed the entire state. There are additional environmental benefits, he adds: "If you can get farms near the city to produce food sustainably, it would enhance air and water quality and reduce erosion where most people live." He's referring to the pesticide and fertilizer runoff that seeps into riverways throughout the Midwest, a result primarily of industrial agriculture. It all flows into the Mississippi River and ends up in the Gulf of Mexico, creating a periodic oxygen-depleted dead zone that in 2008 was the size of Massachusetts and is destroying the northern Gulf shrimp and fishing industries, worth $2.8 billion annually.[16]

Food systems expert Mike Hamm's numbers square with Henry Brockman's. A St. Louis native, he holds the C. S. Mott Chair in Sustainable Agriculture at Michigan State University. Much of his work focuses on farming in Michigan. In a talk given in July 2009 in Grand Rapids, Mike stressed that sustainable food systems can be an economic driver in a region struggling with a deteriorating auto industry but blessed with abundant water and farmland. This is a strength relative to the rest of the world, but also domestically. Currently 50 percent of Michigan's fruits and vegetables come from California, and yet, according to the American Farmland Trust, 86 percent of that land is under threat of development. Moreover, California is facing a serious water crisis that is worsening with climate change: snowpack runoff that feeds the region's water supply could decline by 70 percent over the next thirty to forty years.

Alarming as California's trouble is, it offers a great opportunity for Michigan's farmers, whose land is also endangered by sprawl. Mike's studies show that if the 10 million people of Michigan doubled their current consumption of fresh, frozen, or canned fruits and vegetables—which would simply meet USDA minimal nutrition standards—and if even 15 percent of that food were grown locally, it would require moving 37,000 acres (currently devoted to corn, soy, wheat, and dry beans) into fruit and vegetable production. That alone would generate an increase of $200 million in farm income for the state and almost 1,800 new off-farm jobs.[17] Bear in mind that these figures apply only to current growing practices in wintry Michigan, a state with a brief growing season. Farmers could gain even more income by constructing unheated hoop houses—inexpensive, temporary greenhouses made of PVC pipe, wood, and plastic—in which they can grow thirty different crops year-round. A 3,000-square-foot high-tunnel hoop house can generate $8,000 in gross annual income. A new farmer just starting out on 3 to 5 acres of uninherited land (which goes for some $3,000 an acre in Michigan) could pay back the investment in a couple of years with just one of these contraptions.

These figures speak to the economic self-interest of farmers. From the perspective of the larger economy—and particularly for smaller cities located amid rural areas—sustainable practices keep the flow of farm income within local communities. A study of southeastern Minnesota dramatizes how much money leaves local economies under the current system. It shows that between 1997 and 2003, local farmers sold an annual average of $912 million into the global commodity market, but they spent $996 million, at a net loss of $84 million each year (covered by federal subsidies and off-farm work). If that's not striking enough, the region spends $1 billion a year on farm inputs and food produced outside the region. Both individual farmers and the local economy are losing money in the commodity system, while urban population centers struggle to find fresh, nutritionally balanced food.[18]

I had lunch with Mike Hamm in September 2009 to learn about all this and to ask him in particular how the development of sustainable agriculture might bear on the future of smaller industrial cities. "The point of locally integrated food systems," he says," is that they're scalable—locally, regionally, and nationally. The first choice should always be local." Mike is a middle-aged white man of quiet intensity who came of age working against racial discrimination in Detroit in the 1970s. His professional reserve relaxes as we talk. I wanted to know where that 100-mile rule so scrupulously observed by locavores came from. "That's totally arbitrary," he says, with a trace of annoyance. "It comes from a book called *Plenty*, written by a couple of kids whose hearts were in the right place, but . . ." He shakes his head. I also asked what he thought of vertical farming—the integration of food cultivation into buildings and other tall platforms intended for densely settled cities. "Oh, that's just stupid," he says flatly. "It will do nothing to reduce the carbon footprint. There are plenty of other places to grow food without that." As new urbanists would cry, with horror, *transect violation*![19]

Mike sits on the Michigan Food Policy Council, established by Governor Jennifer Granholm in 2005. Food policy councils, which also exist on municipal, county, and regional levels, are growing in number and fast becoming instrumental to reestablishing what they call "food security." Since governments don't plan for food security as they do for other basic needs such as water management and transportation, these councils argue that vast inequities in food access have emerged as part of the larger urban crisis. Just as farmers have had difficulty staying out of the commodities business, city dwellers in poor neighborhoods—and, in the Rust Belt, entire cities—have seen their grocery stores disappear. The purpose of food policy councils is to find ways of cutting through the array of government and private programs that deal with food issues in a fragmentary way, often at cross-purposes, and to restore the relationship between farmers and consumers in locally integrated food systems.[20] For farmers who want to wean themselves from commodity agriculture, that means providing training in sustainable cultivation as well as programs to help them gain access to land, credit, and liability insurance. It also entails developing local food processing facilities and brand labeling ("Select Michigan") to help consumers identify locally grown fruits and vegetables.

Above all, food policy councils help establish large, reliable markets for locally grown food. One increasingly common way to do that is through institutional marketing, pioneered by the California-based Farm to School Network. Originally the program induced public schools to buy a share of the ingredients used to prepare school lunches from local farmers rather than contracting only with big suppliers serving highly processed food. From modest beginnings in the mid-1990s, Farm to School programs had spread from 400 in 2004 to more than 2,000 in forty states in 2009, with funding from the USDA. Since then the institutional marketing strategy has been replicated to include colleges and universities, hospitals, and senior housing. It has also been promoted by famed chef Alice Waters, whose legendary upscale locavore restaurant in Berkeley, California, Chez Panisse, and several books on the virtues of local food have lent the movement a bit of celebrity cachet.

More prosaically, institutional marketing is securing an ever deeper foothold in places like Michigan, and often in connection with larger regional goals. Traverse City's Michigan Land Use Institute, nationally hailed for its 2003 success in coordinating a six-county partnership to prevent the construction of a highway and bridge south of the city, is developing its regional food system as part of a long-term "Grand Vision" project to curtail sprawl. In 2004, the institute piloted an early Farm to School program under the direction of Patty Cantrell, who sits on the Michigan Food Policy Council. It now provides about a dozen locally grown products, from apples and winter squash to eggs and meat, to more than thirty schools in the region. In 2008, the state altered its funding requirements to take advantage of new federal programs to subsidize school lunches that include locally grown food. Seventy-three percent of Michigan's public school food service directors have expressed interest in clearing the many hurdles to doing so. Meanwhile, Patty told me during a visit to Traverse City, "conservative local business interests have begun to see the entrepreneurial value of farming: it's no longer viewed as a lost cause."[21]

Michigan local food programs are also trying to reach the working poor. By the end of 2009, according to the *New York Times*, one in eight Americans and one-quarter of American children were receiving food stamps from the Supplemental Nutritional Assistance Program (SNAP).[22] To meet this need, statewide efforts have been under way to equip farmers' markets—whose numbers grew by 13 percent between 2008 and 2009 and stood at 271 in 2010, the fourth highest in the country—with the means to process SNAP payments. With a promotional grant from the USDA, Michigan is trying to keep pace with the demand: between 2006 and 2010, the number of farmers' markets accepting SNAP payments grew from just three to fifty-seven. This not only makes nutritious food more available to the urban poor; it also represents a growing market for regional farmers, since Michigan cities (like most others elsewhere) severely restrict the for-profit sale of food grown within municipal boundaries.[23]

In his Grand Rapids presentation, Mike Hamm talked a bit about the farmers of the future and the programs available through the Michigan State extension school to help them. In addition to attracting those who grew up on farms and helping them to stay, extension programs are attracting immigrants with agricultural backgrounds, such as the Hmong and Sudanese, who are increasingly going into market farming. The fastest-growing number of new farms in Michigan are owned by Hispanics, he observed. And then there were young people going into farming for the first time and unemployed autoworkers seeking alternative employment or a means of supplementing other forms of low-paying work during these rough economic times in Michigan.

Before the rise of grocery store chains after World War I, small-market farming, or truck farming, appealed to working people, particularly European immigrants and southern African American migrants, who brought their horticultural skills with them. Filmmaker Nancy Rosin, who produced a film documentary on the history of the farmers' market in Rochester, New York, explains in a book accompanying the film that Italian and Eastern European immigrants first grew fruit and vegetables on the city lots where they lived and later, with the rise of automotive travel, grew much larger quantities of produce along the urban edge. Area market farms were concentrated in the town of Irondequoit, less than 10 miles from Rochester's downtown farmers' market. A sizable number of these farmers, she says, had industrial jobs with firms such as Kodak, Rochester's largest employer, and became known as "Kodak farmers." By midcentury, Nancy writes, Irondequoit "had the largest square footage of greenhouse glass in the world to support the demand for fresh food in a climate with long, cold winters." But by the early 1960s, Irondequoit was fast being paved over, making way for postwar housing, highways, and strip malls. In 1963 the once-powerful Irondequoit Grange closed and later became the House of Guitars. The land available for farming is now pushed farther out, but it's still there, undeveloped, only 15 miles from the city.[24]

Driving out of Detroit one day—on a horrific below-grade eight-lane superhighway reminiscent of *The Matrix Reloaded*—I was stuck by a pattern I'd noticed in Rochester and elsewhere: while smaller industrial cities had, like Detroit, lost population to their growing suburban rings, their suburbs didn't seem nearly as dense or as wide proportionate to their size. That would make sense given their longstanding low-market value. I contacted several researchers who might be able to confirm my impression, which speaks to the suitability of agriculture in smaller cities' suburban fringe areas. Not surprisingly, none of them had an answer: Census figures don't distinguish between large and small cities, making comparative studies difficult. The census makes a distinction between micropolitan and metropolitan statistical areas, but the former consists of urban settlements with an urban core population of between 10,000 and 50,000; cities above that standard are folded together for statistical purposes. My own inexpert crunching of 2000 Census data bears out the notion that smaller cities' suburbs are less densely settled and closer in, but that doesn't take into account what economic geographers call "land cover"—from forest, grass, and water to commercial development, roads, and housing—which provides some sense of land potentially available for farming.[25]

Assuming my impressions and those of others I've talked with are accurate, the more sparsely developed, more proximate, and often highly fertile land surrounding smaller industrial cities could be preserved for a revival of market farming in an approximation of Ebenezer Howard's garden city idea. Several recent studies show that in the words of one, "closer proximity to urban consumers," combined with new marketing arrangements, favors farmers who "adapt their agricultural operations to higher value or specialty crops, such as fruits and vegetables." Compared with both recreational farming and traditional commodity agriculture, small, adaptive farms using intensive production methods have the best chance of surviving in metro areas, according to the authors of the study "Development at the Urban Fringe and Beyond." Small adaptive farms are generally between 10 and 50 acres and have annual sales of between $10,000 and $250,000, or at least $500 per acre. In 2000 they accounted for 9 to 12 percent of metro-area farm acreage. They are better able to accommodate the haphazard, unplanned popcorn development on a city's outskirts, and their presence helps to control it. Surveys of urban dwellers show that in increasing numbers, they value living near open and working land. They're also increasingly aware that sprawling residential development costs more in public services: $1.24 for every tax dollar versus 38 cents for infrastructure required by farmland or open space. Situated near urban markets, adaptive farming also offers opportunity for off-farm employment and access to part-time farm labor. "The counterintuitive result" of all these considerations, write the authors of the study, "is that as urbanization proceeds, acres devoted to vegetable production may actually increase."[26]

Popcorn Farming

Virginia (Vicky) Ranney is an old hand at agricultural urbanism—what some are now calling the new ruralism—for she pioneered a new style of housing development, suited to the urban fringe, that integrates agriculture and ecological preservation into its design. In 1992 Vicky cofounded Prairie Crossing, then billed as a "conservation community," located 40 miles northwest of Chicago in exurban Grayslake, Illinois. It includes an organic farm and nature conservancy within its contours. Originally conceived in the late 1980s without an agriculture component by landscape architect Randall Arendt, conservation subdivisions build the same number of houses on a parcel of land but cluster them at higher density levels, leaving the balance free of development for all the residents to enjoy. The economic appeal for developers is that they don't have to reduce the volume of housing, yet clustering brings down the cost of water, road, and utility infrastructure. The idea is not without critics. While part of the purpose of conservation communities is to reduce sprawl, they are usually built anew on greenfields. In addition, many of these developments to date, some of which are gated communities, consist of enormous McMansion-style homes of no environmental value and do not make provisions for affordable housing. Another criticism, by Andrés Duany, is that they perpetuate the appeal of sprawl by "aestheticizing" it. In this sense, they can be, he says, "nefariously anti-urban." Andrés, an old friend of Vicky, was originally skeptical of Prairie Crossing. He now argues for including agriculture—"correctly allocated"—along the transect and views Prairie Crossing as a pioneering venture in agricultural urbanism.[27]

The rising interest in local food and sustainable agriculture, combined with the deteriorating state of shoddily constructed and increasingly abandoned suburban malls, has created an opportunity to reinvent the suburb—both its form and its ethic of galloping consumption. The cheaper the real estate and the less densely settled these areas are, as in smaller industrial metro areas, the greater the opportunity suburbanites have to shape their destinies along new lines. New ruralist ventures are particularly suited to the Northeast and Midwest, where farmland is most greatly imperiled by sprawl. Prairie Crossing, while itself very upscale, offers a compelling early model of how the suburbs can be remade for a spectrum of uses.

"It is not a gated community!" Vicky Ranney told me emphatically one bright windy day in November 2009 as we toured the development. Vicky is endearingly reticent and soft-spoken, a quietly enthusiastic woman in her sixties who comes across more like a cultured librarian than a developer. (Indeed she has served as an editor of the Frederick Law Olmsted Papers and wrote a book on Olmsted's work in Chicago.) With Prairie Crossing, she and her husband, George—a high-profile lawyer, former steel magnate, and head of Chicago Metropolis 2020—have created a place of beauty and integrity. They also had the means and connections to make it happen. Designed by a top-flight team of architects, including new urbanist Peter Calthorpe, its 359 single-family houses, rendered in vernacular midwestern-craftsman style and color, cost about 20 percent more than homes of similar size in affluent Grayslake. Back in the 1980s, when a developer almost succeeded in laying down a traditional subdivision on farmland adjacent to George Ranney's uncle's farm, they together bought the land and attracted investors for the new community. It now sits on a 677-acre site, surrounded by 3,200 acres (including Ranney's uncle's farm) put into conservation easement that enlarges the adjacent Liberty Prairie Reserve.[28]

More relevant to those who might want to replicate the Prairie Crossing experiment on a more modest scale is the list of ten principles to which it is committed. These include not only environmental protection, but also commitment to "a healthy lifestyle" (including sustainable agriculture), access to mass transit, energy conservation, and economic and racial diversity. To these ends, the community, only 60 percent of which is given over to residential space on lots ranging from one-tenth to a half-acre, is host to a 40-acre organic farm, including a flock of free-range chickens (which makes about $300,000 a year), and devotes 25 acres to a farmer incubator program. There's also space for community gardens. All the houses are 50 percent more energy efficient than similar stand-alone dwellings, and a wind turbine on the property supplies a share of electricity to the residents and the farm. A charter school on the grounds, fully integrated with Grayslake's public school system, is heated with geothermal technology. And its restored wetlands, prairie grass, and historic hedgerows have been incorporated into a swale-based stormwater filtration system that empties into a lake and two ponds, which is more efficient and less expensive to maintain than traditional systems. All of this is available for the larger community's use through 10 miles of nature and biking trails ribboned throughout the nature reserve. Anyone can lease the neighborhood's enormous, meticulously restored barn for weddings and other events: this is not an insular enclave. In recent years, Prairie Crossing has also made good on its commitment to principles of mixed-use, transit-oriented development. It is located adjacent to two intersecting Chicago commuter-train lines—hence the word *crossing* in its name—and across from one of the stops, the developers have built a town center: Station Square. The hamlet-like spot consists of thirty-six moderately sized condo units sitting atop a series of commercial spaces hosting an independent restaurant, a printing company, and several shops.

All this is very high end, to be sure. "We're probably weakest on racial and economic diversity," George Ranney told researchers from the Urban Land Institute.[29] Station Square was intended to offer lower-cost housing, but it's still quite pricey: in February 2010, a three-bedroom, three-bath condo was on the market for $449,000. In retrospect, he says, they should have considered partnering with a nonprofit development corporation so as to provide more affordable housing.

That's just one of many ways the Prairie Crossing prototype could be modified to various scales, including infill projects in both urban and suburban settings. Up the road some 80 miles from Prairie Crossing, in Madison, Wisconsin, another smaller development, Troy Gardens, started as a community garden on vacant state-owned land. It is now an agriculturally based community under conservation easement with thirty homes, twenty of which are affordable with leasing mechanisms in place to keep them so. On the west side of Janesville in what the city is calling its greenbelt, another conservation project, the Fisher Creek Neighborhood, is in the early stages of planning. Developed by the owners of a 70-acre family farm, it is currently selling quarter-acre lots for five houses and later plans to build affordable housing with community gardens and space for light commercial enterprises. The Urban Land Institute's Ed McMahon estimates that in 2010, some 200 projects with an agricultural component were under development across the United States.[30]

These projects are hard to put together. They require unconventional funding partnerships, rezoning, and special easements that go against the grain of long-standing planning for functionally segregated uses, new legal instruments for protecting farmland that can't always compete with the land's market value, and negotiation with several levels of government. Often the land lies in more than one municipality, requiring sticky tax-sharing arrangements. They also face resistance from politicians and developers based on the long-established assumption that commercial sprawl is good for the bottom line. And then there's the not-in-my-backyard problem, which can kick up with special ferocity when barnyard sounds and smells are at issue. It's no wonder that new urbanists long preferred to focus on stand-alone greenfield projects where they can maintain control. But it can be done, often with greater profit, by developers willing to risk building at higher densities than the standard subdivision. Smaller industrial metro areas, where sprawling popcorn development is taking shape closer to the urban core, would be wise to encourage such risk taking—perhaps in creative land use partnerships with small adaptive market farms—while leaving rural areas free of development all together.

Retail Cannibals

In 1980, when I was living in Rochester, New York, I participated in a lawsuit seeking to halt the construction of the area's first gigantic regional mall in suburban Henrietta: Marketplace Mall—hence the caption "Mall Mauler" that appeared with a brief local TV news interview I gave (which still makes me laugh). It was the last of a series of lawsuits and administrative hearings, pursued by the City of Rochester, city residents, and the owners of South Town Plaza, a nearby shopping center, targeting the proposed mall's developer, Wilmorite, and the Town of Henrietta. Most of the litigation was brought under the relatively new National Environmental Policy Act of 1969 (NEPA), the interpretation of which was still being developed in the courts and in state and federal agencies. Our predecessor legal challengers had staved off the project for a year and a half, proving that Wilmorite had not provided an adequate environmental impact statement (EIS) of the site, rich in creeks and wetlands; the law was clear enough on this point, and the U.S. Army Corps of Engineers was given time to conduct the study. More ambiguous was whether EISs were required to address a project's economic impact on the larger urban region. The City of Rochester, which stood to lose downtown retail and tax base to Marketplace Mall, put up a valiant fight on these grounds, but eventually gave up in 1979. After the mall broke ground, my fellow plaintiffs and I were able to persuade the federal district court that the new EIS was still inadequate and tried once again to draw out NEPA's intention with regard to economic "urban impact." As a city resident, I spoke to the future mall's adverse effects on urban retail, jobs, housing, and schools—the city's general quality of life. Our efforts didn't get very far, and in spite of the battles won in this string of lawsuits, we lost the larger war—miserably. NEPA was rendered powerless to address the "economic impact" of suburban sprawl on cities.[31]

Although retail began moving out of cities at a brisk pace with postwar suburbanization, it's easy to forget that the full-on exodus didn't take place until the 1970s, with the building of ever more grandiose enclosed malls. As dismal as downtown Rochester became quite suddenly after Marketplace Mall opened its doors in 1982, it actually fared a bit better than did other small-to-midsize cities after the predictable flight of downtown retail during this fateful period. That's because Rochester was home to the country's first modern urban mall, Midtown Plaza, opened in 1962 and designed by architect Victor Gruen. Famous as the originator of the enclosed shopping center, Gruen conceived Midtown Plaza as a mixed-used competitor to the strip plazas and ugly parking lots cropping up in the adjacent suburbs. In a novel arrangement, Midtown provided plenty of parking in a below-ground three-level garage. Anchored by two department stores and a large supermarket, it also included an auditorium and a sidewalk café; above the entire structure stood seventeen stories of office space, crowned by a hotel and restaurant. Gruen's civic vision of pedestrian-centered urban vitality was attenuated, to be sure, since Midtown was a private space controlled by its owners. But it managed to stave off, however briefly, the worst effects of "those bastard developments," those "functional ghettoes," as Gruen bitterly characterized places like Marketplace Mall. It even helped sustain a few nearby small businesses for a while. (My favorite was an old dining-room-style restaurant, The Cosmopolitan, that had been around since the late nineteenth century.) But Midtown Plaza grew increasingly seedy, even as its annual holiday extravaganza, complete with Santa and a weirdly futuristic 1964 World's Fair–style kiddie monorail, continued to draw suburbanites into the city long after downtown had become a husk. Midtown's doom was sealed when its three struggling anchors finally closed in the mid-1990s; it was demolished in 2008. It wasn't a bad run, as malls go.[32]

Meanwhile, Marketplace Mall has morphed into a retail empire. Even as the original structure has grown, it is dwarfed by a woozy sea of big-box stores, ever expanding parking lots, and several cookie-cutter condo complexes built by the same developer, Wilmorite. It's considered a success; another mall, in nearby Ironequoit, is now officially "dead." A tatty assemblage, its enormous, cheap modular shells stand empty, stripped of the chirpy signs that once announced the consumer fantasmagoria that beckoned within, victims of both Internet retail and yet more malls and big-box stores built farther out on the exurban frontier.

What happened in Rochester captures in microcosm a story that has been repeated with dull monotony in American cities large and small, and even in some small towns. As the suburbs have expanded inexorably into the exurbs, they've left in their wake a growing number of hulking ghost malls and big-box stores—the detritus of sprawl. Even before the 2008 recession took hold, some 20 percent of the 2,000 largest malls in the United States were failing. Since 2006, only one has been built, and in April 2009, the country's second-largest mall operator—General Growth Properties, Inc., a name surely suggestive of its metaphysical ambitions—declared bankruptcy.

This seemingly endless cycle of self-cannibalization was foreordained by a 1954 federal tax code change permitting accelerated depreciation of commercial and manufacturing property. Intended as an antirecessionary, construction-trade-boosting measure after the postwar housing boom subsided, it allowed commercial developers to defer taxes for the first seven years of a building's life (it had previously been forty years and was intended primarily for capital-intensive manufacturing structures). It offered a huge incentive to build shoddily and temporarily on cheap land far from urban settlements, with an incentive to move farther out when the tax bill came due. And the consequences were devastating. The measure essentially deskilled the construction trades. It reinforced white flight and racist perceptions of increasingly tax- and job-poor dysfunctional cities. And since only new development qualified for the tax write-off, independent mom-and-pop businesses and older commercial buildings soon found themselves competing—fatally—with now ubiquitous chains such as McDonald's and Ramada Inns, as well as with office parks that moved jobs out of the city. This enormous federal subsidy, whose development lobbyists worked hand in glove with the highway lobby, also encouraged localities to offer tax abatements, economic development grants, and infrastructure support for new building. None of these ancillary measures changed when the tax shelter came to an end in 1986. And, not surprisingly, it led to corrupt dealings between developers and local politicians. The biggest beneficiaries of these real estate and highway subsidies were shopping mall developers: by the late 1970s some 22,000 suburban shopping centers had been built; by the late 1990s, that figure had almost doubled. Inevitably some of them would fail, particularly in the older, first-ring suburbs. For their part, many big-box stores ("power centers" as they're called in the industry) control the land leases and deeds to their properties long after they've vacated them for newer, bigger digs down the road: they'd rather pay the mortgage to leave them standing empty than risk any chance that a competitor might make use of the site, making a mockery of the free market and consumer choice. Clearly the United States has been "overretailed," as architect Ellen Burnham-Jones puts it, allotting almost 20 square feet of commercial space per person, versus 2 square feet in Europe.[33]

This disgraceful history did not play out uniformly, however, as predictable as the exurban feedback loop has become. Once again, urban size and national region matter. As it turns out, the oldest forsaken mall landscapes lie in the Northeast and Midwest, since that is where most of the earliest malls were built—that is, before the rise of the Sun Belt in the 1970s. (Particularly in the western states, whose urban cores were never well developed, malls have some of the cultural cachet once carried by downtowns in older industrial cities.) In addition, the earliest big-box stores, WalMart and Kmart, were centered in these regions, particularly the Midwest. Arkansas-based WalMart initially catered to the rural market in the South and Midwest, seeking to reach people in small towns and cities far from the "shopping opportunities" offered in big urban centers; Michigan-based Kmart focused on larger towns and small cities in the Northeast and Midwest. Many of these properties paved over wetlands with impunity, even after the Environmental Protection Agency came into existence in 1970.[34]

What all this means is that suburban areas of smaller industrial cities are poised for dramatic reinvention. Not as densely settled as big metro areas, they nonetheless have more empty retail space relative to their size. The blighted consequences of these ill-fated land use decisions, however dismal now, offer a range of opportunities for reshaping their entire metro regions.

Resurrecting Dead Malls

Urban analysts are only now taking stock of how suburban shopping malls might be repurposed to accommodate a growing population in a low-carbon future. Urban planning expert Chris Nelson projects that as many as 2.8 million acres of suburban commercial space will be available for redevelopment by 2015 and could provide space for two-thirds of the country's anticipated population growth by 2040.[35] Yet Ellen Burnham-Jones and June Williamson's *Retrofitting Suburbia*, published in late 2009, is the first book to take the measure of the relatively few suburban revitalization projects to date.

How might dead malls and big-box stores be repurposed in ways that prepare the suburbs for a sustainable future? Here, small metro areas face many of the same issues and choices as large ones do. As it stands, some communities (mainly in smaller cities in the South and Midwest) have shown remarkable ingenuity in reusing these sites. In her 2008 book *Big Box Reuse*, artist Julia Christensen documents big boxes that have been turned into new uses such as an elementary charter school, an indoor raceway, a Head Start facility, and a county library. Some are taking advantage of low leasing costs for old big-boxes and the recession-driven 5 percent uptick in the resale trade, typically handled by nonprofits like the Salvation Army. Since 2008, entrepreneur Bob Ticehurst has opened five for-profit Used Book Superstores in suburban Boston stretching as far north as the small city of Nashua, New Hampshire, where his fifth opened in 2010 in an old Tweeter Center building.[36]

Undoubtedly idiosyncratic innovations such as these will continue. But a larger vision of sustainability will require yet more, and such development could go in three directions—not at all mutually exclusive. The most common is to raze the carcass—to "demall," as the industry says—and build mixed-use "town centers" (not to be confused with "lifestyle" centers, which are all retail) on their enormous footprints. As described in *Retrofitting Suburbia*, the best known of these, Belmar in Lakewood, Colorado, a suburb of Denver, incorporates retail, condo, rental, and town home housing, as well as office space, into an open town square layout. The complex is built upward, two to four stories, which increases density in a building form appropriate to the suburban transect zone—something that will be necessary in order to contain sprawl as population increases. It is located on several bus routes, to help limit driving, traffic, and parking; in its previous incarnation as a mall, there was no bus access. Established in 2004, it is also staged to grow in increments that will enable the developer to fund each build-out. A substantial drawback, however, is that like all other town centers, Belmar's streets are not part of the public realm—with all its First Amendment protections—since they, along with the rest of the development, are in private hands, owned by the developer.[37]

A similar, much smaller incrementally planned project is under way in Cuyahoga Falls, an adjacent suburb of Akron, Ohio, but it hasn't been easy. Developer Bob Stark hopes to replace a 25-acre dead shopping center—now demolished and remediated with state and federal tax-increment financing—with a mixed-use town center called Portage Crossing, which will sit alongside a walking and bike path that extends into downtown Akron. Akron mayor Don Plusquellic is an early and energetic advocate of smart growth and new urbanism who, for twenty-two years, has focused the city's attention on the economic value of sustainability as a long-term draw to the metropolitan area. The bike path his administration coordinated is built along an old canal towpath and is intended to bind the suburbs more closely to the city—and on a midday, midweek visit in October 2009, I could see that it was in heavy use. Pusquellic has been out front in building attractive, infill affordable housing in the city, developing a mixed-used community near the University of Akron, and revitalizing the downtown core while preserving its historic architecture. He's even pushing for a highway teardown. Persuading the people of Cuyahoga Falls of the long-term wisdom of sustainable, compact land use planning has been an upward battle, however. Already, Stark's proposal to narrow the four-lane road on which the property sits has met with a hail of community criticism. He's likely to face another huge battle if, when the economy recovers, he requests a mixed-use zoning permit to build 183 planned residential units. Should he want to include plans for community gardens—a small-scale form of agriculture appropriate for the suburbs—it could present another obstacle. He hasn't even suggested it—not yet.[38]

A second approach (and one that Belmar pioneered) is to incorporate green building practices into renovations and rebuilding. Even the International Council of Shopping Centers (ICSC) has discovered the dollar value of green, though mostly as a marketing tool that, at least for now, appeals especially to the affluent.[39] The ICSC worked with the U.S. Green Building Council to establish a LEED-Retail certification standard, still in the pilot stage as of October 2010. It calls for recycling construction debris and using sustainable building materials, making greater use of natural and light-emitting-diode (LED) lighting, installing wind and solar components, and diverting stormwater to nearby aquifers and irrigation systems. It also requires access to public transit and, in a move that reinforces the stereotype of the latte liberal, provides privileged parking for hybrids. More troubling, the ICSC has not taken a position on greenfield development, a continuing blind spot that pervades the industry. Kohl's, for example, has won industry awards for its green building practices, and yet in 2009 the company moved its long-time distribution center in Menomonee Falls, Wisconsin, to a 7.5-acre greenfield site in Ottawa, Illinois.[40]

A third, and more outlandish, approach is to turn these divas of sprawl into handmaidens of sustainability—that is, to return them to agricultural use, put their gargantuan footprints in the service of energy generation, and restore the natural water filtration systems paved over decades ago. As the idea of retrofitting the suburbs gained currency in the late 2000s, a number of venues hosted design competitions for repurposing fallow mall and big-box space. Nearly all of the submissions included one of these three features, most had an agricultural component, and some provided for all three. Washington, DC's Esocoff Architects, for example, drew up a plan for the *Washington Post* that included hydroponic greenhouses, electric car filling stations, a rooftop solar farm, and an orchard planted on part of the former parking lot, along with housing, residential, and retail space. Another, devised by Birmingham, Alabama, architect Forrest Fulton for *Dwell* magazine, included a hydroponic greenhouse from which consumers would pluck vegetables and take them to a restaurant where chefs would cook them to gourmet specifications.[41]

Visions such as these may sound like fevered dreams. They also present nightmarish funding, engineering, and zoning challenges. And they are likely to face fierce resistance. It's hard enough to get zoning variances for backyard chickens and commercial farming on residential property in the suburbs. Suburban culture, after all, is the living embodiment of functional, separate-use Euclid zoning, promising large residential lots, a wide berth for automobiles, and both geographical and psychological distance from the city's minority working poor. After years of metropolitan neglect and crushing economic woes, smaller industrial metro areas are perhaps most resistant to changing these arrangements. It's not even clear whether agricultural mall repurposing makes sense in the long run, given the need to re-densify the suburbs with the intent of preserving outlying farmland for . . . agriculture.

Yet over the two years since the *Washington Post* hosted its design competition, at least two projects for agricultural mall repurposing have come under development. One is taking shape at the Galleria at Eastview Mall in downtown Cleveland. Built in the 1980s with an enormous glass atrium spanning the length of its roof—a style popular at the time—it has dozens of vacancies among its 200 commercial spaces. In early 2010, with a $30,000 grant from the city's Civic Innovation Lab, marketing manager Vicky Poole opened a composting and hydroponics-based farming project making use of the atrium's natural light, called Gardens under Glass. The mall also hosts a farmers' market, and long-term plans call for attracting eco-retail: restaurants, garden supply and health food stores, and shops specializing in goods made with recycled materials. It may be located in the city, but Gardens under Glass is pointing the way toward transforming a suburban architectural form.[42]

Another project, not yet off the ground, has ambitious plans for the countless vacant big-box stores strewn across the national landscape and concentrated in smaller metro areas in the South and Midwest. High-tech entrepreneurial veteran Gene Fredericks is raising capital for a for-profit venture he's calling Big Green Boxes. The model centers on growing fresh greens and herbs through aquaponics—a closed-loop system combining hydroponics and aquaculture in which nutrient-rich fish feces fertilize the plants, and the plants, in turn, clean the water. Using 90 percent less water than soil-based agriculture, Fredericks claims, the process yields ten times the crops. He plans to outfit these big-box warehouses, roomy and already designed for climate control, with carbon-neutral daylighting, rain catchment systems, LED lights, and solar and wind energy, which are more technically and economically feasible than they were ten years ago. Eventually he wants to install raised-bed gardens on the parking lots. The business plan calls for selling to local fresh-produce brokers who supply grocery stores and restaurants, as well as selling crop starts to local growers. Fredericks insists that he won't compete with local farmers but will instead supplement their wares during the off-seasons when brokers are forced to import produce from, say, Chile and Mexico. Although he plans to open his first venture in the temperate San Francisco Bay Area, where he lives, "Our plan is to have Big Green Boxes in communities everywhere," he told *Grist*. "Big Green Boxes' controlled growing environment will have a greater impact in places like Minneapolis and Chicago in the winter, and Las Vegas and Phoenix in the summer." Obviously smaller industrial metros have as much, if not more, to gain from the idea—and so may Fredericks, since their real estate is much cheaper.[43]

Due for a Makeover

"This economic crisis doesn't represent a cycle," General Electric CEO Jeff Immelt memorably observed in 2008. "It represents a reset. It's an emotional, social, economic reset."[44] Immelt was speaking of the role of government policy in shaping markets, particularly in clean energy technology, but his conviction applies with as much force to the way we use land in human settlements. Just as poised for a reset are American suburbs and smaller industrial cities, relatively recent urban forms whose reasons for being are most intimately tied to the modern industrial past.

I've suggested just a few ways that smaller industrial cities' rural and suburban transect zones could hold advantages in local food production—and perhaps, as a result, could have more to gain from reining in exurban development by building upward in their suburbs. As it stands, their abandoned urban cores are also more suited to agriculture than are, say, densely populated New York or Chicago. Why that is the case is the subject of the next chapter.

4
Framing Urban Farming

In her wry 2009 memoir *Farm City: The Education of an Urban Farmer*, Novella Carpenter lends a touch of cultural history to today's agricultural turn. Novella was born in the 1970s to an earnest hippie couple who tried their hands at country living and organic farming. They split up when Novella was only four years old. She attributes the demise of her parents' marriage to rural isolation and loneliness. Learning a lesson from all this, Novella and her auto mechanic boyfriend, Bill, set up their farm—which consists mainly of meat animals and bees—on a forsaken piece of land next to a junkyard near a highway overpass in the struggling city of Oakland, California. To her, one of the main advantages of farming in the city—which has seized the alt-culture imagination in much the same way the back-to-the-land movement did forty years ago—is that it's more sociable. Novella doesn't necessarily like everyone in her community—she confesses to wanting to tell the "yoga people" who chide her for drinking coffee to "saw off their legs"—but she seems to know everyone in her corner of the world, at least by sight. And she's having a blast.[1]

Still, if urban farming seems counterintuitive, that's because in a well-functioning city with a dense, properly proportioned urban fabric, it is. In his book *Green Metropolis*, David Owen argues that New York City is among the greenest human settlements on the planet measured in terms of its carbon footprint. "The average New Yorker," he points out, "annually generates 7.1 tons of greenhouse gases, a lower rate than that of any other American city, and less than 30 percent of the national average."[2] And the beauty of it is that New Yorkers don't even have to try—or to care. Simply by not driving and by living in small quarters stacked in tall buildings, the denizens of Gotham do more for the environment than the most strenuously ecofriendly composter can imagine. He's right, although that's not to say that densely settled cities are unsuited to agriculture of any kind: gardening has long brought joy to the hearts of urban dwellers. Garden plots in low-rise urban residential areas have supplemented working families' food budgets for generations. Rooftop and balcony gardens also have been favored in the city, closer to the urban core. And community gardens, modeled after the victory gardens encouraged by the U.S. Department of Agriculture to supplement World War II–era food rationing, have an honored place in urban culture.[3]

In cities that aren't doing so well, however, and have vast quantities of vacant land to prove it, agriculture on a larger scale—farming—is not only appropriate, but it can be part of a faltering city's salvation. The Northeast and Midwest are filled with such places—"shrinking cities," they've been called or, more felicitously, "cities in transition" and "cities growing smaller." These former industrial boomtowns owe their modern existence in some form to the auto and steel industries. They come in all sizes, and their relative population loss varies. Detroit, Cleveland, and Buffalo are less than half the size they were at the peak of their glory in the 1950s; Chicago lost a quarter of its people and then began growing again in the 1990s. Most small-to-midsize cities in Auto Alley (stretching from Michigan and Indiana through Alabama) and Steel Valley (in Eastern Ohio and Western Pennsylvania) shared the fate of Flint and Youngstown; Toledo, Dayton, Syracuse, and Erie, among others, took devastating hits. Those that lay within the more diversified trade orbit of Chicago, such as Peoria and Rockford, fared better, but still bled jobs and people, as have many of the older mill towns scattered throughout New England—victims of the first wave of outsourcing in the 1920s. All have been further ravaged by suburban flight.

In most of these cities, the new urbanists' caution against "ruralizing" the city is misplaced: enormous stretches of neighborhoods are already rural; block after block is already literally green, with acres of grass standing in mute testimony to the pulsing life that once thrived there. The question is, What to do with this land? A growing number of urban leaders argue that shrinking cities should embrace their smaller size by densifying some neighborhoods and closing down costly city services to others while making it possible for their remaining inhabitants to relocate. Such an effort requires mustering the civic will to dramatically reshape a city's master land use plan and to pass new zoning ordinances that reflect its intent. Given the exigencies of ward politics and the history of racial housing discrimination in older industrial cities, the political challenges are daunting; to date, no one has yet figured out how to see it through. Whether formal "rightsizing" comes to pass, however, it is clear that in shrinking cities, larger-scale agriculture makes a whole lot of sense—at least temporarily. Another approach to vacant land, not at all incompatible with farming, is to restore some of the green infrastructure that was destroyed by the frenzied pace of urban-industrial growth. Bringing nature back into the city in this way would safeguard the water supply and better prepare for the health and safety of future generations. With changes such as these, which modify the new urbanist transect ideal, the Rust Belt could see the birth of a new, if inherently transitional, urban form.[4]

Planners and policy analysts influenced by new urbanist and smart growth ideas, who focus on the fate of older industrial cities, debate these questions with some heat. For a variety of reasons, smaller cities are in a strong position to create this new urban form, and even to lead the way. Yet here too they tend to be overlooked in the urban conversation.[5]

A Tale of Two Hard-Worn Cities

Leaving aside their differing scales, the Michigan cities of Detroit and Flint have much in common geographically, historically, economically, and demographically. Both are in southeastern Michigan, about a two-hour drive apart. Both have been dependent on the automobile industry. Both populations are majority African American. Both are shrinking cities: between 1960 and 2010, Detroit lost more than half its population (1,670,144 to 713,777), and Flint almost half (196,940 to 102,434). And both are situated amid some of the country's richest farmland; before the rise of the auto industry, their economies were predominantly agricultural. So examining them together is like comparing apples with apples—or rather, a large apple with a small apple.

The last time I had visited Detroit and Flint, in the mid-1980s, they were in rough shape. Over the years since, I had heard horrific stories of their further unraveling and seen pictures documenting their descent. But as carefully as I had paid attention, nothing prepared me for the devastation I saw in September 2009 twenty-five years after that last visit. Detroit, especially, was every bit as postapocalyptic as Hurricane Katrina–ravaged New Orleans, but on a much larger geographical scale. Oddly, the downtown cores of both cities looked shinier with new investment. In Detroit, new development—a gleaming medical area, a tech center, two sports stadiums (Comerica Park and Ford Field), riverfront development, and the expansion of Wayne State University—rendered the downtown-to-midtown area almost unrecognizable in a corporate-institutional way.[6] In the neighborhoods, however, it was as though a scattershot bomb had hit them. Once densely settled, if poor, ramshackle neighborhoods were now gap-toothed, with large stretches of unused green space alternating with barely functional houses, boarded-up houses, and a shocking number of burned-out houses—many, many more than twenty-five years ago. This was true even of relatively prosperous neighborhoods, such as the much vaunted Boston-Edison area near the former GM headquarters. In Flint, the neighborhoods had a similar gap-toothed appearance, but they seemed more stabilized: there were fewer uninhabitable houses awaiting teardown. What accounted for the difference?

Replanned with designs drawn up by John Nolen in 1920, Flint is the birthplace of both General Motors and industry recognition of the United Auto Workers, which emerged from the famous sit-down strike at Flint's Fisher Body Plant in 1936–1937. It is also home to an inspired experiment in land banking through the Genesee County Land Bank, founded by county treasurer Dan Kildee in 2002. The land bank has become an essential tool for managing the city's shrinkage. As anyone who has seen Michael Moore's now-classic *Roger and Me* (1989) knows, Flint has been on the skids since GM began closing down operations there in the 1980s. To arrest the resulting housing catastrophe, Dan had championed state legislation, passed in 1999, that eliminated the sale of tax liens to speculators—or "infomercial watchers," as he calls them. These investors profited from tax lien purchases by paying off back taxes on property, thus gaining the right to collect tax reimbursements with substantial interest and additional fees from the owners; after a period of years, they could also acquire the deed, which, in a weak market city like Flint, led to a high volume of dilapidated rental property and housing abandonment, and eventual state ownership—without clear title. Meanwhile, Genesee County was hit with the financial responsibility for maintaining this deteriorating housing stock with a diminishing tax base. The 1999 state law, then, not only curtailed future speculation but made the county "the landlord" upon tax foreclosure, giving county treasurers the power to reduce extra fees and postpone foreclosure in cases of financial hardship.

Dan then took things one step further. He created the land bank in 2002, which gave the county temporary ownership of abandoned property, much of it left to languish before 1999, and control over its future disposition. Land banking, which became state law in 2004, makes it easier to clear the title and stabilizes the property through renovation or demolition, thus making it available for resale. The Genesee County Land Bank also runs an empty-side-lot-transfer program to adjacent neighbors, which puts the property back on the tax rolls. With revenue from taxes, fees, and sales in both Flint and the more prosperous suburbs, the county land bank is self-financing. In the meantime, it allows neighbors to use unsold land temporarily for such things as food and flower gardens. Most of it, however, just sits there awaiting future use. At the time of my visit, 14 percent of the city's parcels were owned by the land bank. Later that fall, Dan merged his land bank policy center, the Genesee Institute, with the National Vacant Properties Campaign to form the Center for Community Progress, with support from the Ford and Charles Stewart Mott foundations, among others. Armed with lessons learned from Flint, it has been instrumental in helping states and local governments develop land banking authority in the wake of the ongoing national foreclosure crisis that began in 2007.[7]

Although the land bank has slowed Flint's free fall, it's not at all clear what the city will look like in the future. Dan is an early and vocal advocate of formal resizing through careful revision of the city's comprehensive land use plan. Flint's last master plan was created in 1965, in anticipation of future growth, and it's way past due for an update. Always a delicate process, the task is made even harder by the loss of the city's planning department in 2002: planning departments are among the first to face the budget axe when times are hard. Many smaller Rust Belt cities are without them, and the recession has led to cutting many more. Like other smaller communities, Flint also lacks what is nearly universal in big metro areas: a regional economic development corporation. So the vacuum is filled by people like Dan (a county official), the neighborhood-based planning commission, area colleges and universities, a local community foundation, and the mayor.

I asked Dan—a tall, teddy bear–like man who often gives in to the temptation to laugh at the ridiculousness of it all—whether he thought Flint's scale might work to its advantage in the struggle for self-reinvention. The city's size makes it "more nimble," he said, but also "more volatile," meaning that the city's smaller population makes it possible to convene the relevant parties—and their constituencies—with less bureaucratic mediation, but also with a greater chance of facing contentious issues directly and an unnerving sense that decision making could be imminent. Biting at the heels of every move is fear and distrust on both the left and the right. Some are still wedded to the old industrial paradigm: they're waiting for new business to come in and put Flint on the path of growth again. ("Not gonna happen," says Dan.) To some, the idea of shrinking the city smacks of government takeover. And then there are those, primarily in the black community, who can still taste the bitterness of 1960s-era urban renewal policies that drove Route 475 through the heart of the city and threw them out of their homes. They wonder, as do others, whether the land bank is keeping the foreclosed property in trust for wealthy developers who will eventually come in and plow over the interests of Flint's current residents.

The new mayor, Dayne Walling, Dan tells me, was once a research fellow at the Genesee Institute. But now that he's mayor and needs to respond to a variety of constituencies, Walling wants to move away from the term *shrinking*; he wants to talk in terms of "shifting priorities," so as to avoid the rhetoric of "surrender" and alleviate the fears of those whose neighborhoods could face demolition. "There are certainly thousands of residential properties that need to be demolished," Walling told National Public Radio in July 2009 while running for mayor. "I mean, they are not fit for human habitation. But it doesn't follow that the other residents on that block want to move to some other place."[8]

The day after I met with Dan, a meeting was scheduled to talk with the mayor and his staff about "how to have a less shrill public conversation" about land use "rightsizing" and developing a new master plan. More nimble, yes, but also more volatile.

In Flint, the ground is well prepared for agricultural repurposing, thanks to the land-market stabilization work of the land bank. It's a good thing, since the entire city of Flint is a food desert: it has no grocery stores. In 2009, the bank used federal stimulus money to hire an urban agriculture coordinator, Roxanne Adair, an energetic young woman whose parents had come up from the Kentucky coal mines to work at GM. I spent a warm mid-September afternoon with Roxanne and land bank lead planner Christina Kelly, who showed me around the city and took me to two urban farms. On the north side, on East Piper Avenue, they introduced me to Harry Ryan, who had just returned from a memorial service and looked somewhat anomalous garbed in funereal black from head to toe as he posed for pictures before his small cornfield. Sixty-year-old Harry, a rhythm-and-blues musician who speaks in a clipped tenor, also hails from a GM family and worked in the factory himself before getting laid off. He has about 3 acres of land bank property under cultivation. With the help of neighbors, he's growing collard greens, broccoli, kale, turnips and snap peas on eight contiguous lots, and across the street he has planted an orchard of eighteen young fruit trees—apple, cherry, and pear. The project started on just one lot in 2006 as part of a neighborhood cleanup effort. Calling themselves the Green Piper Corps, they had soon cleared enough debris that they could see the sidewalks again, which emboldened them to expand the garden into a farm and to regularly mow the lawns of neighbors who had fallen behind. "What do you do with the food?" I asked him. "We give it away," he said. "They don't like to fuss about it, but there are a lot of hungry people in this neighborhood." Not surprisingly, East Piper has become a comfortable place to live again, an oasis in a desert of unimaginable blight, and troublemakers don't dare touch "Harry's Farm."[9]

Harry's ambitions for sustainability don't stop with urban agriculture, neighborhood caretaking, and lifting community pride. He wants to build a hoop house run on solar and a windmill to power his home. He wants to build a playground. And he wants to find a way of selling the farm's produce for cash. The previous spring, the local Ruth C. Mott Foundation asked him to draw up a wish list for a grant to develop and expand his enterprise. The Mott Foundation, in fact, employs a program coordinator, Erin Caudell, who focuses on urban agriculture and green infrastructure. It's not clear where Harry's project is going, but once again it draws attention to the nimbleness factor: well-established and engaged family foundations can have a proportionately bigger influence in smaller cities where they don't have as many competing philanthropic and policy agendas to contend with as larger cities do. (Here, it's important to note that in our privatized economy, foundations have come to play a stronger role in community development than ever before—for better or worse.)[10]

Roxanne, Christina, and I also visited a beautifully maintained 1.5-acre spread, Harvesting Earth Educational Farm, operated by karate instructors Jacky and Dora King. With a grant, again from the Ruth C. Mott Foundation, the Kings bought the land (which sits across the street from their home and karate studio on the edge of Flint) from the land bank in 2006. They view the farm's mission as an extension of the martial arts training they provide children at the karate school: to further instill self-discipline and respectfulness while equipping them with green job skills. The students learn about natural pest control methods, soil cultivation, construction, and water management. No one was there during my visit, but their handiwork was everywhere in evidence: a huge open garden, an 11,000-square-foot hoop house, a big chicken pen with some thirty hens, several composting bins, a rain cistern, and a beekeeping station. The forty-five kids who had worked on the farm in the summer were paid through grants from several community organizations, including the Mott Foundation and Baker College. With their financial support, the Kings bought more vacant property from the land bank, including 5 acres where they are putting in an orchard and a vineyard to be worked by ex-convicts who are learning horticultural skills. They hope that, while providing local food and employment, their efforts will eventually become economically self-sustaining and even turn a profit as a suitable marketing and distribution system grows.

Six months after my visit, I talked with Roxanne Adair by phone to catch up on Flint's agricultural doings. She had a lot to report. Mayor Walling's office had held three visioning meetings in each of Flint's nine wards, in preparation for developing a master plan. All but one wanted to see zoning for urban farming in their neighborhoods. "After that, for the first time ever," she said, "Walling gave a speech where he used the term *food systems,* and I almost started to cry." With development support from Mott and the Local Initiatives Support Corporation, an independent grocery store called Witherbee's was about to open downtown—the first in thirty years. Roxanne was also consulting with a food distribution network in western Michigan that sells to the Chicago institutional market (such as schools and hospitals), to learn how to bring a food processing center to the city. Its work would involve washing and packing fresh produce, as well as canning and freezing, and creating value-added products such as salsa and pickles. She had a proposal in to the Food and Farmers Association for a mobile processing unit so the city could study how local growers would fare in a larger and more permanent facility. If all these marketing and distribution measures come to fruition, agriculture could become a significant sector of the local economy. It would not be the economic engine that GM once was, but part of a more "diversified portfolio," as the economists say.[11]

As I was about to leave Flint and saying my goodbyes, Roxanne pulled me aside. "There's something else you have to see if you want to understand this place," she said. "With your own eyes." She seemed tired, and I was reluctant to take up any more of her time: it was bordering on twilight. But she insisted. So I clambered into her small pickup truck and complied. We drove about 3 miles west of downtown and then stopped. "This," she said, "is Chevy in the Hole." I'd seen pictures of the now-empty former GM property, and I'd seen it camera-panned (while still partially occupied) in *Roger and Me*. But Roxanne was right: you have to see the quiet 130-acre brownfield site to comprehend its chilling majesty. The buildings are long gone. What's left is acre-upon-acre of level asphalt that eventually plunges all the way down to the Flint River: the U.S. Army Corps of Engineers had not only straightened the river to accommodate GM but left behind an artificial valley that weirdly resembled the below-grade superhighway I'd driven out of Detroit on the day before.[12]

When we got back into town, Roxanne handed me a design proposal for turning Chevy in the Hole into parkland filled with trees and bike paths.[13] Many years earlier, Michael Moore had mocked an adjacent short-lived auto museum/amusement park called AutoWorld, opened briefly in 1984, which seemed like an insulting postmodern joke on all the autoworkers who had lost their jobs there. Now that it was gone and the city was growing accustomed to its increasingly diminutive size, it made more sense to plan for green infrastructure, with trees that could sequester carbon and grasses that could prevent toxic runoff into the river. By March 2010, Mayor Walling had hired a green city coordinator, who had begun a composting program on the site, saving the city $300,000 a year on the expense of carting away organic waste. As Flint finds its way toward a sustainable future, who knows? Chevy in the Hole might one day even attract housing and commerce.[14]

The predominantly African American city of Detroit is facing the same issues—and more—on a much larger scale. At 139 square miles, the Motor City is enormous: its footprint could fit Boston, Manhattan, and San Francisco and still have land to spare. In 2009, an estimated 40 square miles stood vacant, a figure that could rise to as high as 80 square miles if the population stabilizes at around 600,000, as projected. Unlike other industrial and commercial cities its size, nearly all of its housing, built for autoworkers with families, is single residential—there are few apartment buildings—so when a home is abandoned, the entire property is abandoned. The fiscally straitened city has been unable to keep up with the rate of demolition needed to stabilize the catastrophe, and it opened a land bank only in late 2009: countless speculators have already scooped up property that they're just sitting on. With the decline of housing prices since 2008, more Detroiters are moving to the suburbs. Meanwhile, the metro area is a classic urban doughnut of 4.4 million people, and growing. Oakland County, to the north, is among the richest large counties in the United States, whereas more than a third of Detroiters live in poverty. In 2009 conservative estimates put Detroit's unemployment rate at roughly 30 percent.

The food situation in Detroit is dire. Not counting the neighborhoods adjacent to the suburbs, the entire city is a food desert. There have been no grocery store chains within city limits for years, even in the relatively stable Midtown neighborhood—home to Wayne State University and the Detroit Institute of the Arts—which, while I was there, had just been rebuffed by Trader Joe's. About 80 percent of Detroiters buy their food at convenience stores run mostly by Chaldeans, a Christian Iraqi community that has lived in the city for many generations and whose stores are flashpoints for racial tension. Food security is a serious, dangerous matter in Detroit.

Enterprising citizens have stepped in to fill the void, making Motown the most promising, yet most complex, template for American urban agriculture experiments to date. Projects range from Earthworks Farm, established in 1997 to support the Capuchin Soup Kitchen, and situated on twenty city lots of "urban prairie" adjacent to downtown, to 2-acre D-Town Farm, set up in 2006 by the Black Food Security Coalition, located on city parkland on the west side. Established in 2003, the Detroit Garden Resource Program Collaborative (whose member organizations include Earthworks Farm and the Michigan State University Extension School) distributes seeds and provides training in soil management and crop cultivation, beekeeping, orchard building, composting, and tree planting through various faith communities and the local schools. The group also provides on-the-job training and summer employment to teens through Green Corps, which in 2007 opened its doors to adults as well. Since 2008, the county treasurer's office has allowed the national Detroit-based nonprofit Urban Farming to clear and grow produce on tax-foreclosed vacant properties. Then there is the Detroit Agriculture Network, which has coordinated hundreds of smaller efforts and cultivated markets for their goods. In all, more than 900 community and small-plot gardens, and a handful of 2- to 10-acre farms were under cultivation in 2009, providing about 120 tons of food, or 10 to 15 percent of the city's food supply. These efforts, and more, have led many to argue that Detroit could become, in the words of journalist Mark Dowie, "America's first modern agrarian metropolis."[15]

That's a tall order, however—even if it were desirable. Above all, the political obstacles are daunting. After more than six years of scandal-and-corruption-fueled turmoil and near-paralysis in the mayor's office, Detroit began formally to come to terms with its shrinkage only in late 2009. Newly elected former auto parts manufacturer and NBA basketball star Mayor Dave Bing committed the city to the idea of "rightsizing" through a revision of Detroit's Master Land Use Plan, last updated in 1992 when the city was still clinging to a model of industrial growth. Bing has a lot to work with. In the breach, in 2008, several nonprofits and local community development groups brought in the American Institute of Architects to perform a sustainable design assessment team study, which is influencing Bing's thinking about rightsizing. Its report, *Greener, Leaner Detroit*, proposes concentrating the city's population in a mosaic of urban villages based on commercial and residential neighborhoods that are still intact, connected by light rail, and turning what's left over to zoning for agriculture, reforestation, and parkland. (The plan also calls for diversifying its engineering and industrial base through "renewable energy technology and manufacturing."[16] Indeed, the short shrift given to the auto industry is striking.) Still, downsizing Detroit is a hard enough political sell, and greening the city an even harder and more ambitious one. After all, Detroit has been the poster child not only for deindustrialization but also for the twin disasters of urban renewal and public school breakdown. Soon after winning office, Bing began the process of slashing municipal jobs, cutting the public school system by half, tearing down 10,000 dilapidated buildings, and reducing bus routes and other services to minimally populated areas. All of this, and much more, must take place within a historical context of deep class and racial division and distrust—both within the city and between the city and the surrounding suburbs.[17]

That history is playing out in the urban agricultural community along three fault lines. One position is represented by soft-spoken former Black Panther Malik Yakini, owner of the Black Star Community Book Store and founder of the Detroit Black Community Food Security Network, which operates D-Town Farm. Malik, who walked me around D-Town Farm one warm September day in 2009, believes African Americans should be leading Detroit's food security movement, and that the jobs and skills development in the emerging agricultural sector should go to the predominantly black population. (The economic stakes are quite substantial, he says: an acre of intensively cultivated land can produce as much as $4,000 in profit.) Malik is troubled by how leadership in the local food movement has been assumed by so many "white women," equipped with university training and grant money. The city has such a strong stake in its agricultural future that, at the time of my visit, the planning department was working with Malik and others to forge a Detroit food policy council from the tension of their differences.[18]

Both groups have trouble with another vision: multimillionaire John Hantz, of Hantz Financial Services, wants to commercialize urban farming on a much larger scale. In fact, his ambition is to build the world's largest urban farm in Detroit, on more than 5,000 acres, combining state-of-the-art food cultivation techniques and renewable energy harvesting. Hantz has already sunk $30 million into the project, mainly by quietly purchasing an undisclosed amount of property for the project, which he argues will provide good jobs and attract a tourist trade to the city. "This is like buying a penthouse in New York in 1940," he has been quoted as saying. "No one should be able to afford to do this ever again." Plans were under way in spring 2010 to develop the first phase of Hantz Farms on 30 acres near the city's farmers' market. To proceed further, Hantz insists that the city will have to zone his property for a lower agricultural tax rate—which many in the overtaxed city, especially urban farming pioneers, take as further evidence of a land grab.[19]

Mayor Bing is receptive to Hantz's proposal as one of many "pilot projects" related to "the greening of Detroit," but he is hardly "on board" with Hantz, as the press insists on repeating.[20] Indeed, here is yet another challenge with which Detroit must contend: because of its size and the sheer scale of its deterioration, it is now at the center of national media attention, forced to carry the symbolic weight of the 2008 national economic collapse. Soon after the federal bailout of GM and Chrysler, *Time* magazine ran an October 2009 special report on its cover: "The Tragedy of Detroit: How a Great City Fell—and How It Can Rise Again." The magazine had purchased one of Motown's really cheap houses to set up a special bureau to cover the ongoing story and was soon followed by the CNN/Money blog charged with doing the same.[21] The city's exceedingly local and painfully delicate negotiations are now laid open to boilerplate media opinion. The Hantz odyssey offers a good example. Moving from the already erroneous claim that the city was preparing "to turn over as much as 10,000 acres to John Hantz to farm," *Fast Company*'s Greg Lindsay wrote a tirade against "starving" the city of services and turning it into farmland, a "model city for locavores." Better to encourage more density, he argues, quoting Jane Jacobs, by diversifying Detroit's automotive monoculture and drawing on the area's R&D talent and craft skill to develop new industries such as renewables. Detroit's survival, he says with heavy sarcasm, will not "be well-served by either shrinking the city's footprint or growing vegetables."[22]

Such arguments are straw men. They distort through caricature and sleight-of-hand. Obviously Detroit will have to do much more than shrink and grow vegetables to save itself, and it is. Among other projects, the city built a Tech Town complex that in 2005 became home to an alternative-energy business accelerator and training facility, the NextEnergy Center. It is also working hand in glove with the state's renewable energy renaissance zone program to repurpose its engineering, battery, and machine tool expertise for the emerging low-carbon auto industry. In fact, resizing Detroit's footprint may be the only fiscally responsible course, since the city's bond rating was reduced to junk status in 2009. Its wealthy suburban neighbors must step up too: Macomb County's bond rating was lowered in fall 2009, putting the region on notice that it continues to sprawl while ignoring its metro doughnut hole at its fiscal peril.

Greening Detroit could be part of a pragmatic postindustrial strategy for attracting new investment and jobs to the city. It could also lead to denser, more energy-efficient housing, including more apartment buildings. But reshaping the city's land use to better serve its citizens, farming on some scale, building a serviceable mass transit system, and planning for more parks and improved water infrastructure will be difficult in the extreme. The last thing the city needs is a national press clouding the debates Detroiters themselves must have about these issues.

Fewer national organizations, media and otherwise, have a dog in the race for Flint's survival, and that could work to its benefit. And where Detroit faces the monumental task of reconfiguring the metropolis into connected "urban villages," a concept that owes as much to Ebenezer Howard's garden city idea as it does to Herbert Gans, who originated the term, Flint as a whole is already closer to the scale of an urban village. That leaves its citizens freer to focus on the city's neighborhoods in a more granular fashion. It could be that while all eyes are on Detroit, Flint—more nimble, yet more volatile—might just surprise us with a truly new hybrid urban form.[23]

Youngstown Moxie

Ron Eiselstein was shocked to hear that the director of the community development office in Youngstown, Ohio, suggested that I call. "They hate me," he said, with a mix of jubilance and bitterness in his voice. A former landlord who defies the city to collect on back property taxes and water bills, Ron was convicted of a first-degree misdemeanor in 2006 for harvesting timber on city property without a permit. A proud free-market entrepreneur, he breezily throws around the word *socialist* when talking about government. When the *Wall Street Journal* profiled him as a potential swing voter during the 2008 election, he was leaning toward the McCain-Palin ticket.[24] Now out of the landlord business but still a big real estate investor, Ron has other irons in the fire, including a venture to raise shrimp in Youngstown. That's why I was sent his way.

I met with Ron at one of his many other projects, a small, folksy retail venue called The Village Pantry in upscale suburban Poland—the kind of place that sells expensive jams and chocolate, oven mittens and tea. His nervous energy was untamed by this nest of nostalgia. After regaling me with stories about his transcontinental background—he's part Filipino, part Jewish American with Ohio roots—Ron talked about his shrimp farming ambitions. As it turns out, the Buckeye State has just the right climate for an emerging freshwater shrimp industry (or rather, Malaysian prawn), and Ron is one of many seeking to capture some of the $8 billion Americans spend each year on seafood imports, a result of the dead zones that make the coastal United States increasingly unfit for fishing and shrimping. Shrimp farms can earn from $2,000 to $5,000 an acre, with relatively little work and a short growing season. In spite of legal conflicts and mutual distrust, the city granted Ron an agricultural variance to pilot-test a shrimp pond (and catch basin) on an acre of land zoned for residential uses but where nobody lives.[25]

Entrepreneurial projects like Ron Eiselstein's will be needed in this former steel town, which has lost some 60 percent of its population since 1950.[26] That's because Youngstown is home to a brave and farsighted land use experiment called Youngstown 2010, the first comprehensive master plan in the United States to embrace shrinkage as a reality and to do so with the aim of becoming a "cleaner, greener, more efficient city." The plan (which drew from similar experiments in shrinking cities in the former East German Republic) was launched in 2005 by the city's then-thirty-five-year-old mayor, Jay Williams—the first African American to hold the post and the city's former community development director—in close consultation with Youngstown State University and various neighborhood groups. Its original green elements were a little vague, but its vision of "smart decline" received huge national attention—Youngstown 2010 won a prestigious 2007 American Planning Association award—and it became the conceptual basis for planning now under way in Flint, Detroit, and some twenty other older industrial cities. Youngstown 2010 broke the back of urban planning models based on perpetual growth. It remains to be seen whether the city can follow through with new zoning ordinances that include not only concentrating its neighborhoods—which involves closing down sewage, plowing, and other costly services to parts of the city to the benefit of others— but also allowing urban agriculture, green space for parks, bike paths, wetlands restoration, and brownfields reuse.[27]

When I visited Youngstown in October 2009, the city was a wreck. Although it had already spent upward of $3 million to demolish 1,200 dilapidated and burned-out houses, many were still standing—many more than I saw in Flint.[28] Community planner Ian Beniston, who had had a hand in Youngstown 2010 since working as a student intern for the city, took me on a tour that included a part of the city that had been platted out for development in the original master plan, dating back to the early

1950s when Youngstown was expected to grow. It was one of the strangest urban settings I'd ever seen: a traditional street layout of quarter-acre lots, complete with street signs, consisting of nothing but trees. Ron Eiselstein had set up his shrimp farm here. With a combination of pride and irony, Ian also took me to an outmoded intersection that still had a traffic light: all four corner blocks consisted of nothing but grass. Based on the logic of resizing, it was a sign of progress that so much demolition had taken place, but we couldn't help but laugh at the futility of that traffic light, swinging in the breeze. Many other neighborhoods were filled, street after street, with what had once stood at that intersection: hovels unfit for human habitation. We also looked at a few small community gardens, and I learned about the fledging farmers' market and local community-supported agriculture (CSA) programs, but I saw nothing like the urban farms sprouting up in Flint or Detroit. No wonder the city had sent me to visit Ron Eiselstein.

Youngstown and Flint are approaching the project of resizing from opposite directions, as it turns out, working with the tools at hand. In Flint, the land bank was crucial to stabilizing the chaotic disintegration of its housing stock and razing deteriorated buildings. Gradually the residents had put the empty lots to use, mainly for agriculture, and only now is the city beginning to revise its comprehensive plan to reflect—and accommodate—the new reality. Youngstown began with a plan, but it has had difficulty reversing the blight: without a state land bank law until mid-2010, the city and county had been unable to agree on plans to address the problem. In fact, as the year 2010 was drawing to a close, the buzz among city watchers was that the great Youngstown experiment might have stalled. If anything, the recession and foreclosure crisis, which hit Ohio especially hard, made it even more difficult for Youngstown to get ahead of its housing trouble, much less to attract investment. "Traditional community development," which focuses on building affordable housing through tax credits, "doesn't work here," Ian tells me. The city can't even get a clear sense of who owns title to its abandoned housing.[29]

Without a land bank, the funds to retain a full-time city planner, or a community development corporation, and with 90 percent of its abandoned property tax delinquent, the city has had to cobble together the "organizational and civic infrastructure," Ian says, to take Youngstown 2010 to the next level: devising a new zoning ordinance that the majority of Youngstown's citizens can live with. With support from the local Wean Family Foundation, which has acted as a stand-in for a community development partnership nonprofit, two groups were getting the ball rolling in 2009: the Mahoning Valley Organizing Collaborative (MVOC), which launched in 2007, and the Youngstown Neighborhood Development Corporation (YNDC), which got off the ground in 2010. Ian, the twenty-six-year-old son of a Youngstown steelworker and a nurse, had just been hired from MVOC to serve as assistant director for YNDC.

As we took a drive out to the south-side Idora neighborhood—or what was left of it—Ian explained their novel approach to reshaping the city. Instead of running pilot projects scattered throughout the city—an eco-neighborhood here, a community garden there, a commercial development over there—the city was targeting specific neighborhoods for all aspects of revitalization and urging citizens to move to them with a $5,000 down-payment-assistance program (a sum that goes a long way in Youngstown), funded by the Wean Foundation. Idora, nestled along beautiful 2,530-acre Mill Creek Park a little more than 2 miles from downtown, is the first of three such targeted neighborhoods. It was a little spooky. The neighborhood takes its name from the Idora Amusement Park, which, as was common practice in the 1900s, was built at the end of a trolley line to encourage development at the city's edge. The amusement park had burned down years ago, and the lease was held by a Pentecostal church with longstanding, never-realized plans to build a church complex called "City of God." Ian and others had worked hard with the Idora neighborhood association to rid the area of liquor-selling convenience stores—hot spots for the drug trade and violence—and in 2010 they were working on concentrating the neighborhood, where, for example, adjacent to one fully occupied street was another with just two shambling houses. They had brought in the 4-H club to create container gardens on the neighborhood's vacant land and were also hashing out plans to open up an entire block for urban farming. Plans were also under way to turn over part of Mill Creek Park to native plants that would help restore the local ecosystem.[30]

As the year 2010 dawned, urban agriculture seemed to be gaining traction in Youngstown. Just six months after we'd met in October 2009, Elsa Higby, executive director of Grow Youngstown, could rattle off a slew of new developments. In just a year, the group's CSA program had grown from 36 participants and three farms to 125 participants and five farms, one of which is within city limits. In partnership with YNDC, 50 of the 125 members will be subsidized. Grow Youngstown was also developing support for a tri-county food policy council, with the aim of keeping agricultural dollars in the region. A bio-ag leadership council had formed to coordinate food and biotech production as job-producing industries for northeast Ohio. And prospects looked good for a state Environmental Protection Agency grant to fund a major composting facility that would provide topsoil for abandoned properties with soil contamination. The denizens of Youngstown had learned all about the fine art of composting from the master—MacArthur "genius" award recipient Will Allen, of Milwaukee's Growing Power urban farm—who had been the keynote speaker at the city's second annual Grey to Green Festival, held the previous fall.[31]

In early 2010, YNDC hired a part-time urban agriculture coordinator, Maurice Small, of Cleveland's City Fresh, an inner-city CSA program. One of his first projects was to assist the Rescue Mission of Mahoning Valley to establish a 2-acre urban farm on land given to the mission by the city. He's also coordinating the Idora neighborhood's urban farm—"That's my baby, my joy!" he told me—where the community is planting a fruit and berry orchard and creating several straw-bale greenhouses constructed from recycled local building materials. Like Elsa, Maurice brings a broad regional food systems approach to his work in Youngstown, having helped to establish several faith-based rural farms—Common Ground and Goodness Grows—that market to the city. By 2011, 4 acres were under cultivation, a kitchen incubator to commercialize value-added food products was in place, and four farmers' markets had been established in the city, all of which are part of a rural-urban cross-learning web of agricultural entrepreneurs.[32]

"Still," Elsa told me, "only a handful of people in Youngstown really understand the role urban agriculture can play in reshaping the city as part of its long-range vision for land use. And funders are really just beginning to see the connection." She thinks it will be a year or so before the civic commitment and funding achieve critical mass. "It's important that it be income producing, and therefore a low external-input model," she said, meaning without the use of costly chemical fertilizers and pesticides. "It's equally important that we count the value of nonmonetary gains—food itself—as income," she says, expressing skepticism about the Hantz Farms idea in Detroit. "It's interesting," she said, "but it's an old big-industrial model for producing jobs. It doesn't foster small-time entrepreneurs that are the foundation of any geographically based community."

"Youngstown 2010 was the spark that lit the fire over the past three years," explained community organizer Phil Kidd of MVOC, when I asked him why there was still so much blight, so little evident progress since the comprehensive plan had been launched in 2005. Phil, a thirty-year-old U.S. Army veteran, has settled down a bit since the summer day in 2006 when he took it upon himself to stand in the middle of downtown holding up a sign that read, "Defend Youngstown," which became something of an informal brand for the city. "We've been raising money," he says from his tidy desk at the MVOC offices, "working in the neighborhoods, and settling on our targeted approach. Now things are in place." In 2009, MVOC completed a survey of the city's vacant property—the city didn't even have an empirical grip on the scale of the blight until then—finding that there were 4,500 vacant structures in various states of disrepair and 22,000 vacant lots. All told, 43.7 percent of Youngstown's parcels were vacant.[33]

Meanwhile, Phil says, "the people of Youngstown have had time to grow more receptive to urban agriculture and other green initiatives." When the comprehensive plan was first charted out, it included zoning for what it called green industrial areas—by which it meant simply nonpolluting enterprises with knowledge-industry tenants in mind. Agriculture was barely considered at that point, and it was grouped together with recreational zoning. Both vague scenarios have changed (there is more about green tech in Youngstown in chapter 5), and they reflect a gradual shift of consciousness and sense of self-interest in the emerging low-carbon economy.

Before I left Youngstown, I had lunch at the Lemon Grove, a cavernous downtown café that had opened two months before. Proprietor Jacob Harver, twenty-six years old, had taken advantage of Youngstown 2010's cooperative relationship with YSU, led by the director of campus planning and community partnerships, Hunter Morrison, one of the key architects of the city's comprehensive plan. The 14,000-student predominantly commuter school, which was cut off from downtown, had constructed a new building that opened the campus up to the city and was about to turn a dead-end street into a through road connected to downtown. Meanwhile, the city renovated a former pharmacy building that houses the café and leases it to Jacob for reduced rent. The city also increased the availability of liquor licenses, making it possible for Jacob and other business owners to attract nearby college students (among others) downtown.

It's a comfortable, impressive place. Jacob, a city native, shows his commitment to the Defend Youngstown ethic not only by selling Phil's T-shirts (sporting a social-realist steelworker logo) but also by going local—which has a whole different feel to it in a place like Youngstown than in, say, Alice Waters's Chez Panisse in Berkeley, California, and its multiplying upscale imitators. Jacob personally refurbished the interior almost entirely with wood from a nearby barn and sells regionally produced food, beer, and roasted coffee. He keeps the place, which also sells local art, open until 4:00 a.m., seven days a week, with a full calendar of events—Polish Happy

Hour, belly dancing, Family Game Day, Stitch 'n Bitch, Kinda Blue Night (jazz, blues, and R&B open mic), a regular Monday night open discussion, Appalachian Jamboree, a paint-and-dance party, and the occasional drag show. As a result, the Lemon Grove is usually packed with a culturally diverse, class-stratified clientele, and is multigenerational in a way rarely seen in bars and restaurants in other American cities.

Cultivating localism in Youngstown is not a matter of ideology or consumer choice alone. It involves participating in a revival of the city's economy and culture as a matter of survival. "A true dependency ethic has been in place here for years," Phil Kidd told me, "that translated from its one-industry workplace to the government." It doesn't help either that as Youngstown began to fall, it became a notorious battleground for Cleveland and Pittsburgh organized crime activity that lasted through the 1990s, reinforcing a sense of passivity and doom in the political culture. Phil pointed me to a study of postindustrial mobility patterns showing that most young people who leave older cities stay in the region. In Youngstown's case, they'd been moving not only to the suburbs, but to other regional cities such as Columbus, Cleveland, and Pittsburgh.[34] That, he said, was starting to change. The university had expanded its curriculum to support community development, offering courses in urban sociology and American studies, grant writing, and Web-based community journalism. Not only were enrollments up at the university, but there were signs that graduates were choosing to stay and apply what they had learned to Youngstown's great experiment; some who had left the area for graduate school were moving back. "Young people don't remember what the city was," says Phil. "They bring an as-is mentality to our situation, and they're challenged by it. Plus, the older, parochial, established, suburban mentality of the city's leadership, as well as the state's, is being replaced by a new generation—fast, fluid, quick, and networked rather than hierarchical—which works best and most effectively in smaller cities."

La Finca

Youngstown and Flint have been so broken for so long that they haven't even been able to attract immigrants accustomed to the low standard of living found in their home countries. Yet the surge in immigration over the past twenty years has dramatically altered a host of smaller industrial cities in a variety of ways. This too provides an opening for urban agriculture. Although large metro areas remain the primary recipients of these newcomers, first-generation immigrants have dispersed more widely than ever before, in part because traditional gateway cities such as New York, San Francisco, and Los Angeles became so prohibitively expensive during the same period. One nationwide study undertaken between 2000 and 2005 shows that midsize cities had an average 27 percent rise in new immigrant population at the same time that more traditional gateways registered a 6 percent decline.[35] The majority, predominantly Hispanic, have settled in the South, Southwest, and West, and, in a notable departure from previous patterns, first-ring suburbs: much has been made of these shifts by think tanks and in the press.[36]

What doesn't get nearly as much play is how this overall demographic change has played out in smaller industrial cities. "Newcomers can counterbalance the outward migration of longtime residents in terms of overall numbers," observes one rare study of these cities, "but the pairing of these two trends can cause rapid demographic change." The Hispanic population of Allentown, Pennsylvania, grew from 11.7 percent in 1990 to 34.1 percent in 2006, while the city's total population remained almost steady.[37] The Hispanic population of the historic mill town of Lawrence, Massachusetts, long a landing place for immigrants, stood at 70 percent in 2000. Shifting away from the metropolitan bias that pervades our understanding of immigration, along with just about everything else related to how smaller cities work, the implications of these numbers are striking. "In contrast to immigrant communities in larger cities," another study summarizes, "the immigrant groups in smaller cities are not well organized and are not represented in the community's civic life. Moreover, many of the human service organizations do not have the capacity to address the needs of the new immigrants—for example, hospitals often lack translators, and schools are not prepared to handle the needs of large groups of non-English-speaking students."[38]

One such city is the former paper mill city of Holyoke, Massachusetts, which has pioneered efforts to cultivate urban agriculture among its Hispanic population. In 2000 Hispanics constituted some 41 percent of the whole. Holyoke is somewhat atypical: the vast majority of its Hispanics are Puerto Ricans who originally were drawn to the Connecticut River Valley to work in the tobacco fields in the 1960s. Technically they are migrants, not immigrants—Puerto Rico is a commonwealth of the United States—and they've had two generations to settle into the city's civic culture. Nonetheless, Holyoke shares many smaller cities' troubles in this regard: its population has shrunk by some 9 percent since 1990 (having earlier lost much more than that to the southern textile mills), its non-English-speaking population has grown by at least 8 percent during the same period, and 26 percent of the city's residents live below the poverty line.[39] For all these reasons, experiments in Holyoke offer a peek into how urban agriculture can ground a struggling immigrant community while pointing the city toward a low-carbon future. Central to this work is the nonprofit community agriculture project Nuestras Raíces, or "Our Roots."

The term *postwar Dresden* is too often used to describe the look of urban decay found in older industrial cities. But it is well suited to the predominantly Puerto Rican south side of Holyoke: looming over the neighborhood up on a hill stands an enormous cathedral with crumbling walls whose roof caved in long ago, resembling a bombed-out relic of old European glory. Past an abandoned gas station with weeds growing through the asphalt, and row upon row of abandoned storefronts, sits something like an oasis amid the rubble: Nuestras Raíces's Centro Agrícola. It was my first stop on a drizzly Sunday afternoon visit to Holyoke in summer 2008. It's an older low-slung building on a corner lot, flanked by a greenhouse on one side and a small restaurant, Mi Plaza, on the other; the complex is fronted by a plaza that in good weather acts as an outdoor café for the restaurant (it was temporarily closed at the time of my visit). Inside are a large meeting room, a community kitchen, and a few offices, all of them empty on this muggy, gray summer day. The greenhouse, though, was bustling with visitors, all gamely tended to by thirty-four-year-old Geraldo Ramos, who runs a saltwater aquarium business, Marine Reef Habitat, from the place. His great love is ocean coral, and he grows sixty-eight varieties. Standing amid countless gurgling makeshift tanks built from discarded materials and surrounded by flats for seeds and seedlings, he explains that on his most recent trip to his native Puerto Rico, he was horrified to see that the coral reefs he'd enjoyed snorkeling around in his youth were dying. "It is my life's work to grow coral here, and to replant it in Puerto Rico where the ecosystem needs it," he says. "Now, I'm educating people about how important this is. I love talking about this stuff!"[40]

Centro Agrícola translates as "agricultural center," and that's precisely what it is. Home to the Nuestras Raíces offices, it is the point of intersection and coordination for the community's farm projects, as well as a host of related initiatives that have sprung from them. The organization got its start in 1991 when a group of twenty neighbors began planting community gardens on a block-size vacant lot, dubbed La Finquita, or "Little Farm." Their purpose was not only to grow food but to transmit the agricultural skills they had acquired in Puerto Rico to the younger generation. They soon formed a nonprofit and in 1995 hired Daniel Ross, the son of a labor organizer and a teacher, who had grown up on a farm in western Massachusetts. Then twenty-two years old and fresh out of college, Daniel put his community organizing and writing skills to work raising money for a larger vision of the project, pivoting off its agricultural work, that included jobs and leadership development for at-risk youth, community health and environmental justice initiatives, and economic development. By 2010, Nuestras Raíces had created eleven community gardens farmed by some 140 families and built a 30-acre farm, La Finca (The Farm), 2 miles from downtown on riverside land provided by the land conservation nonprofit Trustees of Reservations. (Four acres were purchased directly by the group.) La Finca raises poultry and small livestock, grows flowers for sale, and provides lots for incubator farms available to aspiring farmers. It had also arranged to sell produce to nearby Dean Vocational College in an institutional marketing arrangement and to open a farm stand. The care that has been placed in La Finca is everywhere evident, even in the driving rain on the day I visited.

Although Nuestras Raíces works primarily with Holyoke's Puerto Rican community, it has widened its net to include newcomers of other backgrounds who have made their way to the area. Again, the idea is to help these immigrants apply agricultural skills honed in their home countries to new conditions in the United States. To this end, they've provided incubator farm lots, leased by the Immigrant and Refugee Agricultural Initiative, to immigrants from Central America, Southeast Asia, and states once part of the former Soviet Union. It's part of a broader effort, coordinated by the Northeast Network of Immigrant Farming Projects, to settle immigrants eventually on their own small farms. As part of the network, Nuestras Raíces staff have provided training and technical assistance to immigrant farming programs in and around smaller cities throughout New England, including Westfield, Springfield, Lowell, and Worcester in Massachusetts, and New London, Connecticut.[41]

Nuestras Raíces has irons in so many fires that its Web site can barely keep up with them all. That's partly because it is now very well funded. Its operating budget in 2007 was $850,000, plus it has capital grants and ancillary support through partner organizations. It has become a truism that in the absence of government support for underserved communities since the 1980s, private foundations large and small have filled the void and have even set the urban agenda. The Kellogg Foundation alone has been an early and generous supporter, providing funding for the development of La Finca, including its large barn. In 2008, Kellogg also funded the Holyoke Food and Fitness Policy Council, in which Nuestras Raíces plays a leading role through the Holyoke Health Center, with $689,000. In 2007, 90 percent of support for Nuestras Raíces came from a combination of private foundations and federal and state funding.[42]

For some, it would be all too easy to criticize this leg up, and in fact it has caused some tension in Holyoke. "Some view our training with skepticism, as a waste of government and grant funding," Daniel told me in a telephone interview in 2010. "This city is still deeply segregated. But disfranchisement underlines everything we do. Building civic pride and engagement is the point." Politically, there are signs that it's working. In a city that for decades has had a relatively large Hispanic population with virtually no political representation, the year 2009 marked a shift: the city council now had three Latino members out of sixteen, the school board had three members out of twelve, and for the first time, the school superintendent was Latino. "We didn't do any fundraising for them," Daniel says, "but sure, some of our community members did voter registration work through our projects and drove people to the polls." He stresses too that Nuestras Raíces is committed to economic development in skills and communities that have difficulty attracting private investment.[43]

Before we got off the phone Daniel wanted to make sure that I knew about the group's latest endeavor: La Energía, LLC, a for-profit energy services company that will conduct energy audits, provide weatherization upgrades, and install solar hot water systems throughout western New England. Nuestras Raíces is one of three groups participating in this venture in social entrepreneurship; it will receive $540,000 in federal Health and Human Services grants to train workers for La Energía. The company identified a market for energy-efficient services in multifamily housing and small commercial buildings—others had focused on big institutional markets and single-family residential upgrades—and it expects to create twenty-seven living-wage green jobs.[44]

That might sound like a drop in the bucket, but the business will cultivate transferable electrical, construction, and plumbing skills that could lead to significant small business growth in a city the size of Holyoke. If sustainable agriculture, urban and otherwise, can have a disproportionate effect on smaller cities' civic and economic life, the same is true for the emerging renewable energy sector, to which we will now turn.

5

Making Good: Renewables and the Revival of Smaller Industrial Cities

Driving on the back roads of Illinois from Peoria to Rockford one gray November afternoon in 2009, I got a little lost. It's easy to grow disoriented when you're surrounded on all sides by mile upon mile of cornfields. (It's not unlike getting lost in unfamiliar suburban tract neighborhoods.) But I had also been distracted by the rise of windmills over the distant horizon—enormous things reaching more than three hundred feet into the sky. Eventually I spent ten minutes driving through the forest of tall white posts, their blades gently sweeping the sky, before seeing them recede in my rearview mirror. Hungry and needing directions, I stopped at a local convenience store, the largest of six establishments anchoring a tiny farm burgh. "How long have the windmills been here?" I asked the middle-aged clerk and weather-beaten older woman keeping her company. The clerk explained, quietly and a bit warily, that they'd been in place for about a year on land leased from local farmers while her companion glowered at me in stony silence, shooting hate darts as hard as she could "What do you think of them? Of the whole idea of wind energy?" I couldn't help but ask. With that, my would-be antagonist exploded. "We don't get any of it," she spat. "We have to live with these ugly things, but the electricity goes to New York City. We have to live off coal."

A strangely fitting (if grossly disproportionate) parallel was at work here—a shared sense of famine amid plenty. I could find nothing to eat in that store, plopped in the middle of the agricultural heartland, aside from packaged chips and shriveled pink salt rods called "hot dogs." This angry woman felt similarly about the energy economy. We can thank our centralized food and energy systems for both.

Centralization is necessary for getting goods, including energy and food, to huge, densely settled urban markets such as New York and Chicago. But as we've seen in the two previous chapters on relocalizing agriculture along the rural-to-urban transect, there are plenty of ways to supplement concentrated food systems and, in smaller metros, even to supplant them. The same holds true for energy production and transmission.

In this chapter, I will argue that smaller industrial cities can make distinct contributions to—and derive distinct benefits from—a clean energy economy in ways that large cities and small towns cannot. They have land, manufacturing and agricultural skill, waterways, and concentrated urban energy markets of their own stretching across the Northeast and Midwest in a decentralized web that could thrive in a low-carbon future. But it's not easy to make the case. Doing so requires resorting to technical language and concepts that will likely be outdated soon. Low-carbon technology, along with the political will and funding to support it, is shifting at a dizzying pace, with unstable long-term markets that make it difficult to bring to commercial scale. Moreover, this discussion cannot possibly cover the range of pioneering projects that cities of all sizes can pursue to reduce their carbon footprints. In her excellent 2010 book *Emerald Cities*, Joan Fitzgerald has already provided such an overview. With a wealth of case studies, she shows how commitment to low-carbon technologies, pursued in tandem with thoughtful economic development strategies and state policies, can minimize greenhouse gas emissions while saving taxpayers money and providing well-paying green jobs. These include encouraging transit-oriented development and public transportation; outfitting waste management facilities to cogenerate electricity from heat; establishing and increasing state renewable portfolio standards that require utilities to buy a percentage of their energy from renewable sources; creating waste-to-profit networks that divert reusable trash from landfills and make it available to participant organizations seeking cheap, recycled input materials; building on brownfield sites rather than sprawling into open land; and creating workforce development programs to train, equitably, for green jobs. In the shorter term, cities must develop policy standards for new energy-efficient buildings of all sizes and financing models that make it feasible for owners to retrofit buildings with better insulation, lighting, and air circulation and heating systems, not to mention solar and geothermal technologies, without incurring hefty up-front costs. After all, residential and commercial buildings account for 39 percent of U.S. energy consumption and 72 percent of electricity use. Here, I focus on such projects only as they relate specifically to smaller industrial cities as a class, but that doesn't mean they aren't more broadly of great importance and urgency.[1]

It's also important to stress, as Fitzgerald does, that cities can do only so much in the absence of a national industrial policy that cultivates a consistent market for clean-tech innovation. As helpful as the 2009 $787 billion stimulus act was in funding some local clean energy projects (as are projected program budgets in some federal agencies), the United States still takes a piecemeal and grossly inadequate approach to low-carbon industries. We already have to play a game of catch-up in the fields of solar and wind, particularly with China, Germany, Sweden, Spain, and Brazil.

Finally, it remains to be seen how the U.S. auto industry will reinvent itself: What battery technology will it use? Will it employ electrical energy or biofuels, and, if the latter, which kind? Or will it simply produce more lines of small gasoline-and-ethanol-powered cars? Will it be able to regain the market edge it lost to the Japanese thirty years ago? Will it expand to produce more railcars and buses? And will it pay its workers livable wages? The fate of what's left of American manufacturing, still concentrated in small industrial cities, hangs on how all these unknowns play out.

"Oddly enough," Richard Longworth observes in his book *Caught in the Middle*, "many of these small and medium-size manufacturing cities survived the Rust Belt collapse better than the big cities did." Chicago bottomed out in 1980 and has risen again. Detroit, Cleveland, Buffalo, and St. Louis are still in peril. The smaller places played to their own strengths and took measures to stay in play: they had cheap land and skilled farmworkers, offered tax breaks, retooled to compete with the Japanese, and sought foreign investment. Many invested in "meds and eds"—what urban economists call health care and higher education facilities.[2]

Longworth argues that small industrial cities' days are numbered, however ingeniously some of them stitched together an existence during the Rust Belt meltdown. Globalization, he warns, is only in its infancy and will further concentrate the new-era workforce—"the educated knowledge workers, the creative people, the idea-mongers"—in big cities. He assumes, along with the legion of neoliberal knowledge economy advocates who since the early 1990s have forged the prevailing consensus, that U.S. manufacturing is dead and that the United States must expand its research, communications, and financial services sectors for the development of goods that will be produced in foreign lands more cheaply.[3]

That consensus, which has quietly attracted a growing chorus of detractors, faced serious challenge with the recession of 2008. Increasingly it became clear that our economic woes are tied not just to American workers' inability to compete with the low wages paid in developing countries, but also to the financialization of the U.S. economy, which deepened social inequity and led to the biggest recessionary downswing since the Great Depression. The numbers tell the story. In 1950, manufacturing was 29.3 percent of the gross domestic product and financial services 10.9

percent. By 2003, those numbers were almost reversed. Another way of looking at it, this from the Council of Economic Advisers, is that corporate profits from the financial sector grew from 2 to 40 percent between 1964 and 2004. The way to address both structural crises—the loss of manufacturing jobs to cheaper global labor markets and the financialization of the U.S. economy—many now argue, is by addressing a third: our need to replace fossil fuel energy with renewable technologies and greater energy efficiency.[4]

In fact, until it was blocked by congressional Republicans, the fledgling Obama administration began shaping policy to jump-start manufacturing through various low-carbon technologies in the transportation, renewable energy, and building industries. In a variety of ways Obama also signaled his intent to put cities at the center of this vision of low-carbon sustainable development. Not surprisingly, though, his administration has tended conceptually to lump together smaller cities with large metropolitan agglomerations, paying little attention to the distinct attributes of smaller urban scale.

A Clash between Windy Cities

For a brief stretch of time lasting no longer than fifteen years, Muncie, Indiana, took pride in having become a "gasopolis." In 1886 veins of natural gas were discovered in east-central Indiana, and before long, the pokey agricultural county seat was set on the path of industrialization. Since glassmaking required the hot fire kindled by gas, the boom attracted five brothers from the Buffalo, New York, Ball family, who had invented the soon-to-be-famous home-canning jar. Many more factories and mills moved to town, producing iron, rubber, pulp, nails, gears, boots, and shoes. By 1900, the town's population had grown from 6,000 to some 21,000. Just sixty years before, its methods of wheat harvesting and transport had been akin to those of the ancients. Now Muncie had a complex rail system, a board of trade and a chamber of commerce, ricocheting real estate values, and regular visits from big city capitalists, scientists, and newspaper reporters.

Then the fuel ran to a trickle, and the City of Eternal Gas had to reinvent itself from a boom town to a developed industrial city. Immortalized in 1929 as "Middletown" by sociologists Robert and Helen Merrell Lynd, Muncie was taken as representative of the "hundreds of American communities the industrial revolution . . . descended upon," all vying with one another to become "'bigger and better'" cities. Both *Middletown* and its

1937 successor, *Middletown in Transition*, documented the cultural anthropology of "Middletown"—its class and family structure; its civic, educational, and religious practices; its race to get ahead; its encounter with mass culture, the automobile, and modernity—in an effort to understand the character of this new urban form. It's easy to forget, after two thick volumes documenting Muncie's day-to-day life, that it all began with gas.[5]

The city's manufacturing base settled out, and Muncie prospered for most of the twentieth century. Not only did the Ball Jar Company stay and grow, but the city became a center of automotive transmission manufacturing, which relied on a host of local suppliers, from tool-and-die shops to a major steel and wire firm. Most of that is gone now, felled by the decline of the auto industry and the rise of global outsourcing. Indiana Steel and Wire closed in 2003. GM's transmission plant and New Venture Gear shut down in 2006. Transmission and drive-train manufacturer BorgWarner, which had been around in some form since 1901 and had been shipping jobs overseas since the 1960s, finally closed its doors in 2009, taking down with it 780 union jobs—the area's last ones. As for the Ball Corporation, it had expanded into making aluminum cans and then satellite technology, and in 1998 it relocated its headquarters to Boulder, Colorado, to be close to its biggest client, the U.S. Air Force. Muncie was luckier than many other smaller cities: the Ball Family philanthropies endowed what are now the city's two biggest employers, Ball State University and the Ball Memorial Hospital Complex. Coming in third are service industries, such as Wal-Mart and a Sallie Mae debt collection center.

Muncie is now taking steps toward producing renewable energy centered on wind power. In 2008, the Italian gear and drive company Brevini announced plans to relocate its U.S. Brevini Wind headquarters from Chicago to Muncie, where it will produce the giant gearboxes that sit atop industrial windmill towers and harvest the kinetic energy captured by the turbine's spinning blades. Large gearboxes are about the size of an SUV and weigh as much as 30 tons. The Muncie area appealed to Brevini for a number of reasons. As it turns out, the northern half of Indiana has a substantial wind shed, making it a good location for wind farms and thus offering a local market for Brevini's product. In fact, several wind farms are already up and running, and more are planned for the area northwest of Muncie. Brevini was also drawn to Muncie's long, if flagging, history in the transmission and gear business. Windmills consist of some 8,000 parts, most of which lie in their intricate gearboxes, and east-central Indiana is filled with skilled tool-and-die, machining, and gear-cutting laborers and parts suppliers, long employed by the auto industry, who can provide those components. (Brevini may also attract some of its Chicago-area suppliers, too, creating more jobs in Muncie.) Ball State University, which offers degrees in manufacturing engineering and mathematics, was a draw as well. Then there was the push factor: Illinois's high business taxes, not to mention the Chicago area's expensive real estate, made Brevini receptive to Muncie's appeals, working through the regional economic development corporation, Energize ECI. In exchange, Brevini invested $62 million in building a new corporate headquarters and manufacturing plant in suburban Yorktown and anticipated creating 455 jobs. The new firm planned to open in 2011.[6]

As desperate as the Muncie area is for jobs, the deal was not without controversy. The county promised Brevini the construction of a one-and-a-quarter-mile rail spur to move its heavy products and raw materials. But the spur faced stiff resistance from residents of an exurban housing development who feared their peace would be disturbed and their roads closed by the new rail line. Others wondered why the facilities couldn't be built on brownfields with current rail access closer to town, such as the old GM, Indiana Steel and Wire, Delco Battery facilities, or even the old BorgWarner plant. Somewhat surprisingly, Brevini's adamant position that it would be a nonunion shop met with little more than resignation. The company promises to pay on average about $46,000 a year, a figure that factors in compensation for its salaried white-collar staff. The wage for production line workers will likely be closer to $14 an hour, or just above the poverty line for a family of four.[7]

Greater Muncie is slowly creating what economic developers call a business cluster, one in wind-focused renewable energy. Soon after Brevini announced its plans, the German firm VAT Energies and Services decided to open its North American headquarters near the Brevini complex, attracted in part by the rail spur. VAT makes smaller vertical-vane wind turbines and solar and wind-powered streetlights, and provides maintenance services to the wind industry. It will employ more than one hundred workers. Both companies are building their plants to LEED-certified green building standards, adding to the local market for low-carbon products and services. And both will offer workforce training through Ball State and nearby Ivy Tech, with funding through the state, which will have the effect of upgrading both their workers' skills and local educators' knowledge of renewable energy industries. In 2010, another German company was exploring the possibility of building a large wind farm near metro Muncie, which would earn local farmers $7,000 to $9,000 per turbine in lease payments.[8]

Instead of starting the wind industry anew, or "leap-frogging," in the parlance of economic development, Muncie pursued a "transitional" strategy drawing from the skills in gear making already in place. It is also using "linking" strategies to connect unrelated businesses and services to its emerging clean energy industry in higher education, construction, and streetlight services. By early 2011, the wind industry business cluster had established the national Wind Energy Manufacturer's Association, Inc., and opened headquarter offices in downtown Muncie.[9]

Perhaps surprising to some, Muncie's wind energy industry received big support from the state's Republican leadership. Governor Mitch Daniels had replaced the state's commerce department with the public-private Indiana Economic Development Corporation, which provided Brevini and VAT with attractive incentive packages and steered stimulus funding toward building the rail spur. Indiana is a deeply conservative free-market state, particularly in the south, with profound skepticism of anything that smacks of environmentalism and federal regulation. And yet as a result of its success with wind, that attitude may be changing. In 2009, the American Wind Energy Association named Indiana the fastest-growing wind energy state in the country. Upon making the announcement, the association's CEO, Denise Bode, called for "a national Renewable Electricity Standard (RES) to create a long term, U.S.-wide market for capital investment in wind power and spur growth and manufacturing investment in states like Indiana." She also released a poll (though possibly skewed by the firm's Democratic leanings) showing that 81 percent of Indiana voters favored a federal RES requiring electric utility companies across the nation to generate at least 15 percent of their electricity from renewable energy sources by 2021, including 71 percent of the state's Republicans. A 15 percent RES is not as high as strong advocates would like to see, but the poll suggests that the success of wind in Indiana is softening hostility toward two divisive principles linked with liberalism.[10]

The Muncie metro area isn't doing everything right. By allowing its new employers to build on greenfields, it's moving the center of gravity farther from the city while destroying precious farmland. And nothing guarantees that Muncie's workers will be paid a living wage or that its most disadvantaged citizens will be trained and hired in its new clean-tech industry. But along with these profound weaknesses and unknowns, Muncie's fledgling wind cluster highlights several of the strengths that smaller industrial cities can parlay into the emerging renewable energy industry: engineering and manufacturing skill, a significant natural resource (wind), open land for harvesting wind energy, and attractively inexpensive land values relative to stronger-market big cities. If the cluster is a success, it could spin off new American-owned companies in related fields. Brevini itself, a family-owned firm and now a worldwide power transmission concern, came into being in 1960 in the small northern Italian city of Reggio Emilia, whose population has since grown from 70,000 to 170,000.[11]

Chains That You Can't See

In 2008, China became the second biggest automotive producer in the world, replacing the United States and surpassed only by Japan.[12] Among the many reasons for the shift was U.S. automakers' specialization in gas-guzzling light trucks and SUVs. The soaring price of oil the same year, combined with the onset of a massive recession, spelled the industry's doom, leading to a federal bailout and mandatory restructuring of two of the Big Three: GM and Chrysler. A gasp of horror rose across the Midwest and parts of the Northeast in the hundreds of small cities and towns that comprise "Detroit." They make auto parts in long supply chains ending at some forty assembly plants that spit out new cars and trucks. Auto suppliers are divided into four tiers: Tier 4 (raw material), Tier 3 (small parts, such as wire and ball bearings), Tier 2 (small components, such as radiators or ignitions), and Tier 1 (complete modular units, such as seating or suspension systems). When the bottom fell out in 2008, 700,000 people, many of them skilled, worked at these small-to-midsize contracting firms.[13] Because they are so dispersed and thus "invisible," as they say in the supply chain industry, the popular media too often ignore their part of the story.[14]

Until the 1920s, corporate industry followed a pattern of concentration, with the auto industry solidifying the trend with Ford's introduction of the assembly line system. Confined by limited rail and waterway routes, as well as by primitive means of communication, automotive production was centralized in Detroit, and each company centralized all of its functions. Autoworkers toiled in enormous plants (such as Ford's River Rouge Complex in Dearborn) under the watchful eye of corporate headquarters, handling all its big operations from producing engines and complex Tier 1 components like fuel systems to assembling the final product. Smaller inputs were farmed out to the thousands of small shops dotting the Motor City that specialized in tool-and-die making, precision machining, and producing electrical components and chemicals.[15]

In one of American history's great ironies, the flexibility of automotive travel enabled the auto companies to reverse this centralizing pattern and to deconcentrate. That process began in earnest after World War II, when the automakers, especially Ford and GM, began fanning out to both suburban Detroit and smaller cities in other states, mainly in the Midwest. Between 1947 and 1958, they built twenty-five new "runaway shops," as Detroiters called them, in Motown's suburbs alone, and many more in places like Lima, Lorain, and Lordstown, Ohio; Rochester, Syracuse, and Buffalo, New York; and Kokomo and Indianapolis, Indiana. Their suppliers followed. Between 1950 and 1960, the percentage of those employed in the national auto industry in Michigan dropped 16 percent—and that was just the beginning, before the completion of the national highway system. Although the automakers and their suppliers cited Detroit's land pressure and their need to build new one-level plants to replace older multistory facilities, their primary reason for dispersal concerned the control of high-paid, often resistant union labor through both relocation and automation. Predominantly white cities and suburbs, with support from their home states, used more favorable tax policies to lure auto shops, and the practice of "smokestack chasing" kicked into high gear.[16] Detroit, referred to as the "arsenal of democracy" during World War II, also lost out on postwar defense contracts, which went to increasingly powerful Sun Belt states and to California in what amounted to a de facto national industrial policy.[17]

Today much of what is left of American manufacturing is dispersed throughout the small cities and towns of Auto Alley, an enormous swathe of land grouped between the north-south routes of I-65 (from Gary, Indiana, to Mobile, Alabama) and I-75 (between Flint, Michigan, and Atlanta, Georgia). The now-deconcentrated auto industry is geographically divided roughly between the North and South, with Japanese and other foreign transfer companies predominating in the South and U.S. companies in the North. Thanks to what the industry calls just-in-time sourcing since the 1980s—meaning that parts have to be within a short delivery distance from assembly plants—the more than 3,000 parts suppliers in Auto Alley serve both types of firms. As a result, the auto industry has been shielded from the most extreme forms of offshoring that decimated the electrical and consumer goods industries: three-quarters of the parts destined for U.S. auto assembly plants are made in the United States.[18]

The small industrial cities of Auto Alley and elsewhere, obscured by national media attention given to Detroit's troubles, can flourish again in new, more sustainable ways. To do so, their supply shops and engineering infrastructure must draw on their strengths to retool and diversify for the emerging renewable energy economy. Even if the automotive industry transitions into clean-powered vehicle production, experts say that its supply chain is likely to contract in the face of global competition, making it all the more imperative for its suppliers to prepare for renewables.[19]

Consider, for example, the emerging solar panel industry in Toledo, Ohio (known as the Glass City), about 40 miles south of Detroit. Toledo, along with Elmira, in southwestern New York State, has long been an engineering and production center of blown and pressed glass, glassware, fiberglass, and fiber optics. The metro area still provides a share of the windows and windshields that end up in cars and trucks, but it has lost thousands of jobs to offshoring in auto glass and other glass-related industries. Toledo's glassmakers began a transition to solar panels, primarily a glass product in the 1980s, when the University of Toledo opened its Wright Center for Photovoltaics Innovation and Commercialization. Its main innovation has been in thin-film solar, which engineers derived from similar technology used to press ultrathin layers of microscopic material into auto glass to minimize shattering or reduce glare. Its biggest success to date, First Solar, went public in 2006, and by 2009 it was the leading American producer of solar panels, with contracts for huge solar farms throughout the United States and Europe. Its advantage lay in a process using cheaper non-silicon-based raw materials and thus an ability to sell at lower cost. Between 2007 and 2010, its production had quadrupled to 1,282 megawatts, or, by one measure, enough to power about 100,000 U.S. homes. By the end of 2010, its costs were 75 cents per watt, down by more than a third of what they had been in 2006 and well on their way to grid parity, or what it costs to buy traditional forms of energy (coal, natural gas, nuclear, and large hydro) from the electric grid.[20]

In 2007, the university received an influx of state, federal, and industry funding for the Wright Center to coordinate research and incubate new solar businesses. As of 2009, it had spun off seven solar start-ups, including two (Xunlight and Solargystics) that incorporate thin-film solar into building materials. Having gotten its glassmaking foot in the solar door, the Wright Center (which partners with other institutions, including Ohio State and Bowling Green) is now researching advances in polymer-based printable or spray-on nanotechnology that can harvest sunlight and can be produced even more efficiently than thin solar. As of 2009, the Toledo area employed 6,000 people in the solar industry, but it was having difficulty hanging on to manufacturing jobs in solar and other renewables. First Solar has relied on the large German market, but there's potentially a huge untapped market in the United States, too. Although the U.S. market for solar has grown, until it has matured, the industry needs help retooling its manufacturing base. Ohio senator Sherrod Brown's 2009 proposed bill for investing in manufacturing progress and clean technology (IMPACT) would provide a statewide revolving-loan program for small-to-midsize companies to do that, but fiscal restraint has made the future of this legislation unclear.[21]

With a serious national commitment to renewable energy, other smaller cities that contribute to Auto Alley's second- and third-tier levels of the supply chain stand to benefit as well. Akron, for example, is a research and production center for polymers, a chemical advance over plastics. Once known as the rubber capital of the world, Akron is the historic home to the tire industry and wisely made the transition to polymers more than thirty years ago. Recognizing that big tire makers such as Goodyear and Firestone enjoyed an artificial monopoly of synthetic rubber during World War II, Akron's industry and civic leaders planned for a soft landing long before the U.S. auto industry began to flag in the 1970s. Working with the state in the 1980s, they established the Edison Polymer Innovation Corporation to help commercialize polymeric research, and boosted the University of Akron's School of Polymer Science and Engineering and Akron Polymer Training Center. As a result, the Akron area is a world-class polymer center, with 400 firms employing some 35,000 people producing everything from tubing and packaging to liquid crystal display monitors. Because Akron diversified and transitioned early, it's in better shape economically than many other smaller industrial cities—indeed, the Brookings Institution removed Akron from its list of weak market cities in 2007—but its manufacturing base is still struggling. In 1980, 35 percent of the area's jobs were in manufacturing; by 2007, that figure had fallen to 16 percent. And as much as the polymer industry as a whole has been good to Akron, the number of manufacturing workers in plastics and rubber fell a full 50 percent between 2000 and 2007. Meanwhile, the city itself has yet to reverse its more than 25 percent population loss since 1960.[22]

Clearly Akron is in a position to contribute polymer-based manufacturing components to domestic renewable energy industries in solar, wind, biofuels, geothermal, and small hydro, as well as the next generation of sensors and batteries. Renewables could provide markets for many other materials and supply businesses. Youngstown and Buffalo, old steel cities, are home to highly skilled precision manufacturing firms. Rockford, Illinois, once led the world in industrial fasteners, and although the field is in reduced circumstances, it could grow again. Given the instability of the auto industry, some suppliers have already retooled and diversified to handle not only renewables but also such industries as aeronautics and military supplies, which are required by law to contract mainly in the United States and to devote a portion of their work to depressed areas. Michigan suppliers have been particularly adept at this.[23] Over the past several years, Lauren Manufacturing, which employs 200 workers just outside Canton, Ohio, has expanded its polymer sealing and sleeve business, which sells to the auto industry, to include solar sealing and water filtration systems. It helps that as the market for renewables grows in the United States, an increasing number of mature companies, mainly from Europe, are moving their assembly operations to American shores to be close to their final construction sites. Brevini, in Muncie, is just one example among many making this transition.[24]

None of this will translate in any significant way into American manufacturing jobs without a multilayered national industrial policy, one that provides market signals for investment in low-carbon industries willing to open shop in the United States. Senator Brown's proposed IMPACT bill, while important, should be joined with larger federal initiatives, including establishment of national renewable energy and automotive fuel efficiency standards. Meanwhile, if we're serious about creating living-wage American jobs, the United States should reconsider its trade relationship with China. China was brought into the World Trade Organization in 2001 on the assumption, in part, that it would open markets for American manufactured goods. Instead—and quite wisely, from a forward-looking global economy perspective—China has subsidized its export businesses, particularly in clean energy and other green technologies, while manipulating its currency and keeping wages crushingly low, leading to a massive U.S. trade deficit. As a result, neither Wall Street nor Silicon Valley has been willing to resist the market pull toward manufacturing in China products developed in the United States. Yet neither power center is hospitable to American national policies that could balance the trade deficit and employ American workers. Because most among them resist the merest federal regulatory policy in the United States, they take full advantage of China's state power—all while proclaiming the virtues of the free market.[25]

As recently as the 2008 presidential election, conventional wisdom held that American manufacturing was dead and that the United States would prevail by securing its rightful place as a knowledge and innovation driver in the global economy. The tables could be turning, however, and they should. As Ohio economist and auto industry analyst Susan Helper observed in a *Washington Post* op-ed during the heat of the election, "Even the most modern economies cannot thrive without making things. We need manufacturing expertise to cope with events that might present huge technical challenges to our habits of daily living (global warming), leave us unable to buy from abroad (wars) or leave us with nothing to sell that others want."[26]

Given their recent history as links in the automotive supply chain, small industrial cities are poised to gain from a revival of American manufacturing in a low-carbon economy. To flourish, however, they must find ways of working together as never before.

X Is the Silicon Valley of Y

It has become a given among economists that economic clustering, after the fashion of Silicon Valley in the 1990s, gives businesses a competitive advantage in the global economy. The idea here is that innovation springs from synergies generated by the geographical concentration of talent, ideas, and capital. In an age of globalization, markets are respecters of no boundaries—national, state, county, or municipal—and function instead as vast networks that stretch across the world. In this system, political jurisdictions can do little more than impede the fevered search to join new technologies with financing products and end users in the most cost-efficient way possible.[27]

Before the auto industry deconcentrated, Detroit was something of a cluster itself—a gigantic one. When the automakers began outsourcing after World War II, cities and towns throughout the Rust Belt and the South responded with the practice of smokestack chasing, an economic development strategy that put them in competition with one another for chunks of large corporate businesses. When Muncie lured Brevini away from the Chicago area, for example, it had engaged in a bit of artful smokestack chasing. The trouble with the practice is that it's a zero-sum game from a regional perspective, especially in a contracting economy. During the postwar period, cities and states attracted businesses by building office parks and factories on greenfields and directing public funds toward highway infrastructure geared to freight trucking. As one analyst describes the formula, "Recruit a factory to the edge of town, and give away the farm to get it!"[28] Competition grew even more vicious in the early 1980s with the new federalism, when states and localities were expected to attract industry in public-private partnerships with aggressive tax incentives, allowing businesses to drive down costs in the global market. One study of twelve midwestern states shows that a whopping 80 percent of 2010 state economic development budgets alone went to paying for these recruitment incentives.[29] Pressure to accommodate whatever arrangements would create jobs was all the more intense given the increased efficiency of commodity agriculture: off-farm employment has been growing among farm operators over the past fifty years.[30]

Not surprisingly (yet shockingly), it goes unremarked that smokestack chasing is historically and economically tied to the fate of smaller industrial cities as a class. It has deep cultural roots in the Midwest dating back to the real estate boosterism that was so relentlessly mocked by the literary set in the 1920s. But the practice became more intense as these smaller places competed with one another for pieces of "Detroit" after World War II. Today, as large cities such as Chicago have found a secure place in the global network of urban agglomerations, smaller cities still pursue a strategy of smokestack chasing—perhaps all the more fiercely as they are floundering. They also have the most to gain from abandoning the practice in favor of business clustering arrangements that advance regional competition. Since the Rust Belt, with its mixed agricultural and manufacturing legacy, is poised to prosper in a low-carbon economy, it is all the more necessary that they coordinate regional strengths. After all, not everyone can build windmill components.

Describing these complex, multitiered regional economic development arrangements in any depth would take us too far afield. Most have only recently come into being, they are still rare, and by their place-based nature, each is distinctly structured to reflect local strengths, only some of which bend toward clean energy. Generally they share several characteristics, as described by Mark Drabenstott who until recently headed the University of Missouri's Center for Regional Competitiveness.[31] They consist of self-defined regions made up of a group of counties. Their stakeholders usually include some mix of a nonprofit private sector economic development corporation, local universities and colleges with expertise to develop new patents and systems, business incubators that can commercialize innovation and spin off entrepreneurial start-ups, state and local political leaders committed to focusing public investments on targeted regional strengths, and, ideally, regional philanthropies that can provide a neutral safe place to help participants transcend turf protection and forge new partnerships. Drabenstott has been involved with two multipartner regional economic development groups: Riverlands Economic Advantage Partnership, established in 2007 with participants from fourteen counties in the tristate area where Wisconsin, Iowa, and Illinois meet, and the Southern Minnesota Regional Competitiveness Project. Toledo's solar industry benefited from a similar organization, northwest Ohio's Regional Growth Partnership, which began in 2002 as a fairly typical smokestack-chasing outfit and, under new leadership in 2005, became more aggressive as a business accelerator, organizing the area's first venture capital fund (Rocket), with ties to several state clean energy programs. These include Ohio's Third Frontier green funding initiative and the state-funded Wright Center for Photovoltaics Innovation and Commercialization at the University of Toledo—one of eight Wright Centers of Innovation intended to minimize tech-research competition among the state's universities. (The university is also home to one of the state's seven Edison Technology Centers, there devoted to food processing and packaging: Toledo is pursuing several green initiatives.)[32] By 2008, ABC News ran a story on Toledo, with all its "synergies," dubbing it "Solar Valley."[33]

Partly because it has had a well-established regional economic development group working on its behalf for a relatively long time, central New York State is becoming "the Silicon Valley of green technology," as one of the area's politicians proclaimed.[34] That's a bit of hyperbole, yet the Syracuse area can lay claim to a widening array of clean-tech successes. The Metropolitan Development Association of Syracuse and Central New York (MDA) has been around since 1959, when the city was still a thriving center of manufacturing centered on General Electric, Solvay Chemical, the Carrier Corporation (air-conditioning), Crouse-Hinds (traffic signals), and two of the Big Three: Chrysler and GM. While the business leadership organization initially served only metro Syracuse, MDA began partnering regionally in the 1990s and today helps frame economic development for a group of twelve counties that include the small cities of Utica, Ithaca, and Watertown. In 1996 MDA prepared Vision 2010, an unusual step for a small metro area region at the time, which identified seven industry clusters ripe for growth and targeted investment. Among these were indoor environmental quality and energy systems, drawing from the local engineering legacy of Carrier, which had moved its headquarters out of state in 1990 and closed its two manufacturing facilities in 2003. As a result, the area was well positioned to participate early on in emerging green building and energy-efficiency initiatives: the U.S. Green Building Council had been established in 1993 and issued its first LEED standards in 1998. As with Ohio's Edison Centers, New York established a program for strategically targeted research centers, from which Syracuse University won a $15 million grant in 2001 to focus on environmental quality systems. It was the precursor to the establishment of six Centers of Excellence (CoEs), each with a particular focus—from nano electronics in Albany to imaging-based infotonics in Rochester.[35] By 2004, the governor had created the Syracuse Center of Excellence in Environmental Systems, with an expanded mission to include "renewable and clean energy sources" emphasizing biofuels; it later expanded further to include water resources management. In all, the Syracuse CoE has secured more than $72 million in state and federal funds, and in March 2010, it opened a state-of-the-art LEED Platinum building (on the remediated brownfields site of an old typewriter manufacturing building) with office and laboratory space for its many partners and collaborators.[36]

The Syracuse CoE acts as a statewide funnel, bringing together researchers from local universities (and beyond, to some extent), public and private funders, established businesses, business-incubating services, and workforce training projects, all focused on green building, renewable energy, and water resources research. What's supposed to come out of the narrow end of the funnel is a world-renowned research-and-development hub, an environmentally sustainable local ecosystem, and jobs in the low-carbon economy. Here's just a sampling of some of the projects CoE has a hand in. Through CoE's state-funded CleanTech Center, the State University of New York College of Environmental Science and Forestry (SUNY ESF) in Syracuse established a biodiesel production program in 2006, refining methods for turning its cafeteria's waste cooking oil into fuel for the college's vehicle fleet. (The students draw public attention to their project by turning the locally based New York State Fair's annual 900-pound butter statue—a tribute to the state's dairy industry—into 90 gallons of biodiesel.) Engineering faculty from SUNY ESF and Syracuse University won grants to develop a more sophisticated reactor and to turn the glycerol by-product of biodiesel into biodegradable plastic—a renewable polymer that several medical device companies have shown interest in developing.[37]

Among the fourteen start-ups already under way as of this writing is e2e Materials, based on research at Cornell University, which makes glue out of soy resin and has brought out a lightweight, naturally flame-retardant form of particle board based on the product. Company president Patrick Govang is working with MDA to find a local market in animal feed for the soy meal by-product of his crushing process. CoE has also supported prototyping research in efficient heating, cooling, ventilation, and air purification systems, some of which have gone into production. Still others are working on automated water pollution sensors, biomass mapping studies of area farmland, and an energy-efficient computer data center at Syracuse University. The CoE building itself is part of a downtown revitalization effort in the eight-block Near Westside, which is designed along new urbanist principles. The district is showcasing green building practices in its retrofits and is attracting green-tech companies to the neighborhood. Syracuse University is also building a new arts center in the neighborhood, and the entire project is training local workers for green jobs in the building trades.[38]

It's not yet clear to what extent CoE's many green-tech initiatives are translating into jobs. The city itself has not been particularly cooperative in providing a market for green building jobs; it requires only new public construction to be LEED certified, and even then on the relatively low silver level. And the city's premier green building project, longstanding plans to expand Carousel Mall into an enormous tourist-attracting green complex, Destiny ("the world's largest green mall"), is stalled—possibly permanently. Many, of course, are not supportive of the concept, including the leadership of the MDA, which has remained silent about the mall's shaky prospects.[39]

Still, MDA hopes that its green strategy, boosted by an unusual merger with the city's chamber of commerce in 2010, and now known as the CenterState Corporation for Economic Opportunity, will eventually find markets not only regionally but beyond by developing a supply chain of products in demand throughout the world. The credit crunch of the 2008 recession hasn't helped. But the MDA has persevered with other regional issues such as reducing ticket prices for its low-market airline access and developing a one-stop job database for regional employment opportunities. These measures and others sprang from a 2004 economic development plan, The Essential New York Initiative, based on studies conducted by two urban development consulting firms: the Battelle Institute and Richard Florida's Catalytix. Not surprisingly, as a result of their recommendations, the MDA focused on attracting and retaining "young talent" and "transitioning to a knowledge-based economy" with a regional branding strategy, New York's Creative Core, featuring a green apple logo.

While I was interviewing MDA president and CEO Robert Simpson in summer 2009, he received an important phone call. Tall and rangy with a basketball player's build, and scattershot funny, he is, at thirty-three years old, extremely young for someone in his position. He told me excitedly that the call was from a plug-in-car manufacturer. When I queried him, Rob thought MDA would assist the company with automotive R&D testing.[40] Consistent with remarks he'd made at a conference six months earlier when I'd first met him, he was emphatic that manufacturing jobs were not going to return to upstate New York. Yet that fall, Bannon Automotive announced that it had selected the Syracuse area, over locations in Michigan and Kentucky, as its site for assembling a new line of plug-in electric cars. By 2010, Bannon, which has exclusive rights with Reva Electric Car Company of Bangalore, India, to develop the car for the U.S. market, was planning to set up operations at a recently closed Ball Corporation plastic bottling plant in suburban Lysander and expected to employ 250 people making 30,000 cars a year by 2012. The car, the NXR, is an affordable, smaller alternative to the high-end electric vehicle models developed by Fisker and Tesla, luxury electric car start-ups, and expects to price the plug-in for $17,000. Bannon CEO Paul Wilmer made the location decision in large part based on "Central New York's position as a national leader in green technology."[41]

The MDA's long-term clean-tech regional planning, initially centered on indoor environmental systems, has expanded in all kinds of unanticipated ways. Now that support for a low-carbon economy is gradually gaining political traction, the area is poised to capture entrepreneurial "synergies"—as economic developers like to put it—undreamed of just a short time ago, even in manufacturing.

X Is the Saudi Arabia of Y

As the so-called third industrial revolution in renewables unfolds, smaller industrial cities have a manufacturing base that can be retooled to make wind, solar, hydro, and green building components, and the next generations of low-carbon vehicles, best pursued through regional cooperation and business clustering. But they also have natural assets—in land and waterways—for the generation of clean energy. The electric grid, moreover, could be restructured in ways that channel locally harvested energy to serve their smaller urban markets. Rarely, and only partially, are such favored circumstances given a hearing, and they are never discussed as a collective advantage. It's time to make the case.

Renewable energy technology is changing with the speed of light in a desperate bid to compete with the established efficiencies of energy in coal, big hydro, nuclear, and natural gas. All bets are on: by 2007, renewable energy sources (excluding conventional hydro) accounted for the largest portion of added capacity, yet as of early 2008, only 3 percent of the electricity Americans consumed was generated by renewables.[42] The policy landscape too is undergoing massive overhaul. State laws have been in flux since the 1990s, when Congress broke up the "natural monopolies" of the private utility industry, handing states the authority to set terms for restructuring these once vertically integrated generation, transmission, and distribution businesses. Congress debated federal climate change legislation, on and off, over the first two years of the Obama administration. And between the 2008 federal stimulus package and huge increases in its annual budget, the U.S. Department of Energy has received more funding than ever before, leaving Secretary Steven Chu with a blizzard of project proposals that even his expanded agency has had trouble keeping up with.[43] Indeed, the future of the technology and policy framework lies beyond the scope of this book. It is clear, however, that U.S. dependence on electrical energy has grown dramatically with the advent of consumer computer technology, from 10 percent of total American energy consumption in 1940 to 40 percent in 2005. It is likely to increase even more, since, in the auto industry, electric vehicles and plug-in hybrids appear to be winning the race against hydrogen fuel cell technology.[44]

It's widely recognized that the electric grid is in need of upgrading. Essentially the system operates on the same centralized principles designed by Thomas Edison in the 1880s. Energy is harvested and turned into electricity at large generator plants far from population centers; its voltage is increased and its current reduced by a substation transformer for travel over long-distance transmission wires, then greeted by another substation closer to market that adjusts the voltage and current to match supply with demand, and a local utility distributes the electricity to consumers. The system is exceedingly delicate, in part because battery storage capacity is minimal and costly, requiring precise calibration between supply and demand. (A great deal of R&D is going into battery technology for both automotives and the electrical power system.) The system is not designed to accommodate smaller fluctuating loads, and yet currently our two most advanced forms of renewable energy, solar and wind, are highly intermittent: the sun shines only part of the time, and wind currents vary. Both are inadequate for providing consistent base loads of electricity.

From one point of view, then, upgrading the grid requires building more cross-continental ultra-high-voltage transmission lines capable of carrying larger bulk loads of power, mainly wind energy from the Midwest and Great Plains to the heavily populated coasts. Doing so, according to some estimates, would cost something in the neighborhood of $1.5 trillion by 2030.[45] (The emphasis on wind is based on estimates that onshore wind power can produce as much as 37 million megawatts in electricity, which is nine times as much electricity as the United States currently consumes.)[46] Critics argue that constructing a national transmission superhighway is not only too expensive but also endangers energy security by further centralizing the system and builds in energy inefficiencies due to transmission loss, which at long distances runs to about 7 percent. It's also likely to wreak political havoc over establishing rights of way, and in any case, it's unnecessary. "What's needed," writes Kevin Bullis summarizing some of these arguments in MIT's *Technology Review*, "are improved local and regional electricity transmission," along with development of the smart grid and refinement of low-carbon generating technologies, rather than "transmitting power from North Dakota to New York City."[47]

The so-called smart grid, often and confusingly folded in with calls for upgrading the national grid, will be important to communities of all scales seeking access to affordable low-carbon energy. Already constructed on a municipal level in Boulder, Colorado, the smart grid calibrates the transmission and distribution sides of the system through interactive computer automation, enabling suppliers to integrate intermittent wind and solar energy into transmission lines and enabling consumers to match their electricity use with cheaper, off-peak energy flows through time-of-use metering. (This is not to be confused with net metering, which allows consumers with alternative-power home installations to sell electricity back to the grid if they produce more energy than they need.)

It is on the energy generation end of the electrical grid that smaller industrial cities have distinct contributions to make and advantages to snare. Decentralizing energy production is generally cast in terms of distributed generation in amounts ranging from 1 kilowatt to 5 megawatts, in contrast with large-capacity plants generating 500 to 3,000 megawatts.[48] With such a wide spread in generating capacity defining the terms, it's easy to see why debates about distributed generation tend to divide the world between small-scale projects and the voracious energy demands of large, populous cities. In *Green Metropolis*, for example, David Owen argues that it is better to invest in more big power plants to meet the needs of "a central city, like New York" rather than to cultivate distributed generation: because of the city's high density, the energy is used more efficiently. "Part of the fascination with distributed generation," he argues, "arises from our very worst impulses. . . . The desire to produce your own power in your own basement is akin to the desire to drive yourself to work and swim in your own pool and play tennis in your own court: to be liberated from the grid is to be liberated from other people."[49]

While fantasies of techno-dominance and personal liberation have a long-entwined history in American consumer culture, that is a false contrast. It's true that much discussion of distributed generation is given over to residential power installations. Part of the reason for that lies with the success, to date, of residential solar (both photovoltaic and solar thermal), particularly in California, which has among the longest-standing and most stringent state renewable portfolio standards in the country: 33 percent by 2020.[50] Geothermal heating and cooling at the building level has also become increasingly common, and the use of that technology has been expanding to cover multibuilding or district systems as well. In fact, Ball State University in Muncie broke ground in 2009 to replace its aging coal-fired boilers with the country's largest district geothermal system (not to be confused with "big g" geothermal, available only in the western states where steam can be harvested directly from hot rock deep below ground for electricity generation). Ball State is drilling some 4,000 5-inch-diameter holes running 400 feet below ground to make use of the earth's consistent 50 degree temperature in a sophisticated "small-g" closed-loop heat-pump and refrigerant system that will heat and cool about forty of the campus's fifty buildings at a savings of $2 million and 80,000 tons of carbon emissions a year.[51]

It's also true that the literature on low-carbon urban strategies is filled with awed accounts of small towns that have gone off the grid entirely, such as Varese Ligure, Italy (population 2,358), Rockport, Mississippi (population 1,500), and Freiamt, Germany (population 4,400).[52] To Owen's point, it's hard to imagine a densely settled, large metro area powered by a significant proportion of distributed generation in the foreseeable future—if ever. Making their buildings and infrastructure more energy-efficient (along with smart growth transportation policies) makes much more sense, at least for now. To that end, the Clinton Climate Initiative's C40 group, established in 2006, has worked with representatives of the world's forty largest cities to develop best practices and procurement guidelines for measures such as heat and electricity cogeneration, waste management methane harvesting, weatherization, and residential financing incentives that tie up-front costs with the property rather than with the owner (such as Berkeley FIRST, in California). Since 2009, the C40 group has made that information available to all local governments.[53]

But there's plenty of practical middle ground between the solitary consumer-geek "going off the grid" and centralized power generation. While it makes sense from a green perspective for huge metro areas to draw most of their electricity from distant, already existing power plants, smaller industrial cities are in a position to supplement a substantial proportion of their grid energy with locally generated energy organized at the utility level. Besides, smaller industrial cities, particularly in the Midwest, have been loath to support a cap-and-trade system because they are highly dependent on coal and would be penalized in the deal as a result. They have much to gain, however, from the development of a more localized and regional system. They wouldn't have to give up land for new transmission lines carrying power from which they wouldn't benefit, and they have land and topographical resources, proximate to their population centers, suited to renewable energy generation.

Currently there are two feasible ways to bring more locally generated renewable energy online, keeping jobs in the community and electricity at a reasonable cost: through municipal utilities and community choice aggregators, a relatively new form of public power distribution. Many states offer consumers renewable-energy-credit programs for selecting clean energy sources (at a higher price) but no control over how far away their electricity is generated. Here, it is necessary to step back and take account of structural changes in the electrical utility industry in the 1990s. When the Federal Energy Policy Act of 1992 broke up the so-called natural monopolies of the private utility industry, it opened up transmission lines to all generators, making it possible to get more alternative sources online. It also allowed states to require private utilities to divest their generating plants, forcing electricity providers to compete for generators' price on the wholesale market. Many states opted not to deregulate at all, or did so only for a short time (until the California rolling-blackout crisis of 2001), leaving private, or investor-owned, utilities free to own all three generation, transmission, and distribution functions—mostly in the western and mountain states. And while they may be constrained by state-imposed renewable portfolio standards, big private utilities left unregulated offer consumers little choice over where and how their electricity is generated (aside from expensive renewable-energy-credit programs). In some seventeen states, wholesale energy generators compete for buyers on the open market. That, combined with universal access to transmission lines, makes it possible for renewables to get into the game, but who will mediate the terms? It's a particularly vexing question given that renewables are still more expensive, even though the average price came down from $3.48 to $1.75 per kilowatt-hour between 2000 and 2009.[54]

Municipal utilities, as nonprofit government-owned entities, offer the potential for greater public involvement, independence, and technical innovation. These publicly owned utilities, including a few on state and county levels, serve 40 million people and about 16 percent of the market (by contrast, investor-owned utilities serve 71 percent of the market, and the rest are served by federal projects and rural area cooperatives). Because they do not have to pay taxes, handsome private sector executive salaries or quarterly profits to shareholders, and can borrow low-interest bonds, public utilities can sell electricity at lower prices than their private counterparts—on average, between 9 and 13 percent lower in 2009.[55] They can also apply their savings to new projects and infrastructure upgrades and are better positioned to make long-term plans.

With support from voters, municipal utilities in Austin, Texas, and Sacramento, California, have contracted heavily with local solar and wind suppliers, with a large share of citizens willing to pay a premium for clean energy while offsetting their costs with residential generation and home efficiency measures. Denton, Texas, entered into an agreement in 2009 to purchase a full 40 percent of its annual electricity portfolio from a wind farm just 30 miles away. Holyoke, Massachusetts, had the good sense to purchase full ownership of a local hydroelectric plant in 2001, passing cheaper rates on to its customers, and outfitted the city with fiber-optic broadband. As a result, the city attracted a $100 million green supercomputing center to be used for climate modeling and biotechnology development in collaboration with MIT, the University of Massachusetts, Boston University, the Defense Advanced Research Projects Agency (DARPA), the National Science Foundation, and Cisco Systems, among other partners. It was under construction at a downtown location in 2010. And in 2008, Gainesville, Florida, became the first community in the United States to devise a feed-in tariff structure, charging consumers a locked-in solar tax rate of $0.32 per kilowatt-hour for twenty years paid to a fund for further development of renewable generation. The system provides suppliers the kind of market stability for renewables that has brought success to the green energy sector in other countries on a national scale, notably in Germany, Spain, and Denmark.[56]

Municipal utilities are not the exclusive preserve of small cities. They range in service size from small towns to global cities such as Los Angeles, which has deployed its independence and market size to impose a renewable portfolio standard, or RPS, that exceeds the state's—the highest in the country. Los Angeles is also considering a feed-in tariff.[57] Most municipal utilities, however, are concentrated east of the Rocky Mountain states—legacies of the Progressive and New Deal eras—and cities that already have such arrangements are in a position to take advantage of their flexibility in purchasing from renewable sources. Those that don't might consider participating in the growing municipalization movement, though they should be prepared for stiff opposition from private utilities and their political servants.[58]

Some smaller cities and towns dependent on investor-owned utilities can pool their numbers in community choice aggregators (CCAs), which use their bulk purchasing power to negotiate prices and terms with energy suppliers. They have the power to require that a percentage of their electricity is purchased from local renewable energy suppliers. The biggest and perhaps most successful CCA is the Northeast Ohio Public Energy Council, consisting of 126 communities. In 2009 it entered into a nine-year agreement with FirstEnergy of Akron, which will save participating customers an estimated $19 million a year and make available a $12 million grant for use in renewable energy programs in the council's communities. FirstEnergy has a decent renewables track record, with a commitment to meeting Ohio's RPS requirements and with plans to build the country's largest biomass facility in Shadyside, Ohio. It has also invested heavily in regional transmission lines, which suggests that it's planning for local markets.[59]

CCAs could go further and insist that a high proportion of locally produced renewable energy be included in their bulk purchases, at least in theory. But so far, such efforts have been stalled, in part because the relatively high price of alternative energy conflicts with CCAs' appeal as a cost-reducing mechanism. Besides, big private utilities—and they've been growing bigger and more powerful since deregulation—are generally hostile to the idea. A number of communities throughout California (including San Francisco), for example, have been fighting hard since 2002, when CCAs were approved by the state, to institutionalize these bulk purchasing arrangements. They've been particularly vocal about their intent to use CCAs to purchase large amounts of locally produced clean energy. California's largest private utility, PG&E, has put up a fierce fight that has included sponsoring a statewide ballot initiative requiring a two-thirds community majority vote to form a CCA—or to expand a municipal utility—which it lost in June 2010, leaving the way open to CCA success.[60]

As of 2010, CCA-enabling legislation had been passed in only five states (Massachusetts, New Jersey, California, Ohio, and Rhode Island), and only two CCAs had been established, in Ohio and among the communities of Cape Cod: the Cape Light Compact. Smaller industrial cities, many of them located near sources of renewable energy, would be wise to urge CCA legislation in their states and to form these compacts where they already can. Large metros without municipal utilities can also benefit from such arrangements, though due to the size of their markets, they are already in a better position to bargain with investor-owned utilities to increase their supplies of renewable electricity—though sourced from a far distance.[61]

If smaller industrial cities can find ways to distribute low-carbon electricity through locally controlled utilities, they also have, by historical happenstance, land in abundance to generate it. Most urban dwellers are blissfully unaware that producing significant quantities of alternative energy requires land. "A good rule of thumb," a solar industry executive told me in late 2008, "is one megawatt requires about eight acres of land"—and that's in sunny California, where the weather is optimal for solar power.[62] Wind power, unless it's sited offshore, obviously requires large tracts of land that are preferably far from large population centers, due to the noise that windmills make and the hazards of tossed-off ice in cold climates. And by definition, biomass and biogas technologies require farm-and forestland to generate the raw resources required, as well as the physical plant to conduct the conversion.

The Midwest has been called the Saudi Arabia of biomass—which refers to organic matter that can be converted to energy through fermentation—due to its plentiful crop production and animal waste. Smaller industrial cities could reap huge economic benefits from the more efficient development of biofuels, specifically biodiesel and cellulosic ethanol. As it stands, federal subsidies and market incentives for corn ethanol have done little to lessen U.S. dependence on foreign oil or reduce carbon emissions, since it relies on high-carbon fuel for processing and heavy fertilization and pesticide inputs. Corn ethanol has also taken millions of acres out of food and animal feed production while raising the price of land once affordable for smaller farms. And since it harvests only the corn kernel for fermentation, it's not very efficient. The second and third generation of biofuels, only recently in commercial development, could transcend these limitations while bringing down costs to compete with the price of oil, which will only rise. The two principal classes of biofuels, biodiesel (based on vegetable and animal fats) and cellulosic ethanol (based on organic fermentation), are competing with each other, and it's not yet clear which will predominate or whether they might end up in a Mac-PC-style draw. What is clear is that both rely on a greater variety of crops and that cellulosic ethanol has made the biggest technical breakthroughs over the past few years. Cellulosic ethanol makes use of all parts of a plant, including woody matter, and thus can be harvested from corn stover (the stalks, leaves, and cobs that remain in the cornfields after the grain harvest), tree by-products, straw, and prairie switch grass, all of which require less input. The industry is also developing dedicated, high-value "energy crops," as well as new strains of bacteria capable of hastening the fermentation process.[63]

Taking advantage of 2010 federal renewable fuel standards and Department of Energy loan guarantees, biofuel from POET, the world's largest ethanol producer, is transitioning from corn to cellulosic ethanol to produce 3.5 billion gallons of cellulosic by 2022, including a new plant in Emmetsburg, Iowa. The company also has joined with Magellan Midstream Partners to build an 1,800-mile cellulosic ethanol pipeline from the Midwest to New Jersey. The pipeline alone is expected to create up to 50,000 construction jobs and 12,000 permanent positions. Just as important, POET expects to sell biofuel at the pump for $2.00 a gallon by the time the Iowa plant opens in 2012, down from $4.13 a gallon in 2009—and likely competitive with oil prices.[64]

Of course, it's not at all clear what kind of market will exist for biofuels of any kind over the long run, since the automotive industry hasn't yet settled on electric versus biofuel technology, not to mention vehicles powered by hydrogen fuel cells. But whatever way this shakes out, smaller industrial cities can only gain as centers of agricultural production for both forms of power. According to one estimate, the United States exports 58 percent of its wheat, 34 percent of its soybeans, and 18 percent of its corn—all of it heavily subsidized. This practice not only drives down the cost of food in Third World markets, to the detriment of local farmers in distant lands, but also supports multinational shipping concerns, which garner as much as 40 percent of the price for export crops to move them overseas. That money could be going to American farmers and processors, and the local economies they support, while providing home-grown sources of energy.[65]

Biogas is another emerging biomass technology from which smaller industrial cities, surrounded by farmland, can draw in a local energy economy. Appropriate mainly for commercial dairy and swine farms, the process involves harvesting methane, a combustible greenhouse gas, from livestock manure, and converting it to electrical or thermal energy with an anaerobic (or oxygen-free) digester. The digesters thus serve two purposes: they remove an extremely harmful greenhouse gas from the atmosphere and generate a clean form of energy. The energy is used on site, with the excess sold to the local grid. The Crave Brothers Farm of Waterloo, Wisconsin, which produces milk and cheese, provides one model of how to do it. The Crave family's 2,200 cows produce 50,000 gallons of manure and 3,000 of cheese whey refuse each day. In 2006, they partnered with biogas energy systems firm Clear Horizons, which received support from the state's Focus on Energy program. Clear Horizons built the farm's two 750,000 gallon digester tanks; manages them over the Internet from Milwaukee, 60 miles east; and handles the sale of electricity and digester by-products such as potting soil and cow bedding. In 2009, Clear Horizons netted $300,000 in electricity sales to the local electrical utility, WE Energies, generating enough clean energy to power some 450 homes. Meanwhile, the Crave family is free to concentrate on farm operations while receiving low-cost renewable energy as well as the fertilizer by-product.[66] As important as the size of the Crave operation is to the model's success is its location near a heavy distribution wire. As a state agriculture department expert told WisBusiness.com, "If you are more than half a mile from a three-phase (heavy duty) distribution line, or a natural gas line, it's probably not going to be practical economically."[67]

As of April 2010, 151 anaerobic manure digesters were in operation across the United States, most concentrated in big dairy states such as Wisconsin, Pennsylvania, Vermont, and New York, generating almost 400,000 megawatts in electricity or its thermal equivalent annually.[68] Most are not connected to the grid, but many more could come online if local distribution networks were in place to handle their output.

It's easy to forget, too, that smaller industrial cities are almost uniformly located on river ways that can be tapped for hydropower. Established as farm and raw-material-processing settlements before the rise of the railroads, these communities relied on rivers—further connected by canals and the Great Lakes into a grand water transportation system—to get their goods to market. Indeed, the European American settlement of the Midwest coincided with the opening of New York's Erie Canal in the 1826 and the construction of the last significant canals in Illinois in 1848. Gristmills and, in New England, early cotton mills and textile factories also relied on the waterpower provided by high-gradient rivers before the widespread adoption of coal-powered steam by the 1840s.[69] By the twentieth century, with the widespread use of electricity, more of these rivers were dammed up and put in the service of generating electricity (as well as providing for flood control, drinking water, and irrigation). The biggest hydroelectric dams were built in the West and, of course, Tennessee with the enormous federal Tennessee Valley Authority project, but few river systems escaped the hydroelectric-building craze of the mid-twentieth century, not to mention the scourge of industrial pollutants.

Although hydropower is a source of renewable energy, industrial cities have been hard on rivers and the ecosystems they nurture. Some larger systems are still in operation (with improvements), such as the Great Stone Dam (15 megawatts), built in 1848 in the old mill town of Lawrence, Massachusetts, and the Boott Hydroelectric plant (24 megawatts) in nearby Lowell, both on the Merrimack River.[70] One environmentally sound, if rare, way of harvesting hydropower is to outfit non-electricity-generating dams and reservoirs with turbines: of the 80,000 dams in the United States (most operated by federal authorities), only 2,400 are used for hydroelectric power.[71] As much of the smaller infrastructure has fallen into disrepair, however, environmental advocates argue for tearing down some older out-of-service dams in an effort to restore river flow and riparian habitats, from wetlands to fish-spawning grounds. Others (ideally run-of-the-river, or without a reservoir) can be upgraded with more efficient turbines outfitted with fish screens or, more rarely, integrated into commercial renovations of old mill buildings ("micro hydro") as in the Square One development in Holyoke, which generates 500 kilowatts—enough to make the complex energy independent, with electricity to spare for sale to the local grid.[72] Most of these efforts, along with outfitting pumped-storage dams with cogeneration equipment so as to be self-generating, don't produce enough power to make a significant contribution to local grids. They do, however, take users off-line and reduce dependence on fossil fuel.

New in-river hydrokinetic technology, still in the pilot phase in 2010, shows more promise. Kinetic hydropower functions much like wind power but under water, where rotors (or other conversion devices) are turned by the water's motion. Since water is 832 times denser than air and therefore packs much more kinetic energy, and water flow is constant (unlike intermittent wind and solar), hydrokinetics has distinct advantages, and not just in free-flowing rivers. Harnessing energy from ocean waves, currents, and tides holds the biggest potential. Based on proposals in the pipeline in 2007, experts predict that by 2025, hydrokinetics could harvest 13,000 megawatts of power—the equivalent of twenty-two new coal-fired plants or the emissions of 15.6 million cars.[73]

In-river hydrokinetics is almost completely undeveloped. An exhaustive 2004 study of "low head/low power" resources showed that 21,000 megawatts is available for such development, most of it located in the states east of the Mississippi.[74] In 2010, a number of pilot projects were under way using two main technologies. One, already commercialized in the small city of Hastings, Minnesota, on the upper Mississippi River, produces 250 kilowatts of electricity as a downstream adjunct to a run-of-river hydroelectric dam using an underwater, portable barge-mounted system.[75] A far more ambitious project conducted by Massachusetts-based Free Flow Power, farther south on the Mississippi (eighty sites) and Atchafalaya (seventeen sites) rivers, is testing not only barge-suspended technologies but also turbines grounded in riverbeds and attached to bridges.[76] In 2010, the Federal Regulatory Energy Commission issued approximately

100 preliminary permits for hydrokinetic energy projects concentrated in the Mississippi, Missouri, and Ohio rivers.[77] This is low-hanging fruit, testing technologies in some of the nation's swiftest inland currents. It eventually could be feasible in river ways throughout the Northeast and Midwest, feeding electricity to the local grids of small cities along, say, the Susquehanna, Mohawk, Hudson, Connecticut, Merrimack, Rock, Illinois, Wabash, Miami, and Missouri rivers.[78]

While renewable energy technology gets up to commercial scale, the Obama administration has a transitional policy of developing traditional energy industries in nuclear, "clean" coal, and offshore oil drilling. An enormous natural gas field extending from upstate New York through western Pennsylvania and eastern Ohio and West Virginia, is also under development. The Marcellus Basin consists of gas hidden deep within shale rock, requiring hydraulic fracturing, or "fracking," through horizontal drilling, a technology developed only in 2008. Youngstown, Ohio, which lies within the region, won bids to open manufacturing plants to produce the steel tubing necessary to siphon off the gas, creating hundreds of new jobs.[79]

By the spring of 2011, in light of the BP oil rig disaster in the Gulf of Mexico, the nuclear catastrophe at Japan's Fukushima Daiichi power plant, and the release of the film *Gasland* documenting the hazards of fracking to the water supply, it wasn't at all clear how these transitional policies would play out politically. If nothing else, these and other challenges make it clearer than ever before that Youngstown and other smaller industrial cities' brightest economic prospects belong with renewables. Renewable energy, after all, is permanent—another way of saying "sustainable"—and requires stable places to steward its harvesting.

Relocalizing food and energy production in these places, rather than treating their people and natural resources as disposable, could also ease longstanding cultural grievances, like those directed at windmills (and the urban elites they serve) by that woman in rural Illinois. Valuing smaller industrial cities as essential to our future in a decentralized low-carbon economy could even lay the seeds for a renewed sense of common moral purpose.

6

Roots of Knowledge: Local Economics, Urban Scale, and Schooling for Civic Renewal

"They care more about what's going on in Nicaragua than in Springfield just thirty miles down the road," Bob Forrant tells me in a deep grizzled voice over the phone. "Not that there's anything wrong with studying Nicaragua, but still . . ." The former machinist-turned-academic-historian is talking about his graduate studies alma mater and former employer, the University of Massachusetts Amherst. His 2009 book, *Metal Fatigue,* explores the historical anatomy of the metalworking trades in the lower Connecticut River Valley, from its earliest Revolutionary War days making munitions for the Springfield Armory to the closure of the area's last large machine tool firm, American Bosch, in 1986. By the Civil War, Springfield had become the country's "industrial beehive," a center of innovation that cultivated skilled workers in precision manufacturing and forward-thinking entrepreneurs. Together they pioneered new technologies that fueled the second industrial revolution.[1]

Bob worked for twelve years at "the Bosch" and served as the union's business agent until the plant's shuttering. He tells the story of the factory's gradual takeover by outside investors and manufacturing conglomerates seeking ever higher and more rapid returns on their investments, and who disinvested in the local industry while outsourcing jobs. At the time of Bosch's closure, it was making precision fuel-injection systems for cars, trucks, and tanks. But Bosch's days had been numbered since the 1950s, when the auto and defense industries began their exodus from the Northeast, and as the company contracted, no one had the sense to come up with contingency plans for the local economy. It's a ghastly and familiar tale of disinvestment and disposability, which hit small industrial cities like Springfield with special force.

What interested me particularly about Bob's book was its close attention to the Connecticut River Valley's production networks of skilled metalworkers, whose skills had been refined and transmitted through many generations of industrial upheaval since the days of the Springfield Armory. I wanted to know if a critical mass of such skill still existed in the area (as it does in Auto Alley) and whether local economic development leaders were preparing for a possible revival of manufacturing in renewable energy and low-carbon automotives that could put those skills to use. "It's diminished," he tells me, "but it's still very much alive. If you drew a circle with a 50-mile radius around Springfield, you'd find some 300 shops—distributors and suppliers—employing 14,000 to 15,000 people." These fifteen- to twenty-person firms, often family owned, employ toolmakers, metalworkers, and plastic injection molders (along with administrative staff) who have been making parts and machine tools for New England's medical device industry, what's left of the defense-related aerospace sector in the Hartford area, and one-of-a-kind surgical instruments. "Hand them a blueprint, and these people can make prototypes and models for anything, whether it's a windmill turbine part or a golf ball," says Bob. "The trouble is these skills aren't being passed on. These guys are in their mid-sixties, on average, and their kids aren't following them into the business. There are no succession plans. This area once had really excellent programs that trained for today's computer-run machines at places like Westfield Vocational Tech High School and Springfield Tech Community College. Courses in cosmetology, food services, and health care have replaced training in sheet metal and electrical work for training in today's computer numerical control machines. The old training-apprenticeship-educational regime provided an upward ladder for workers, from shop floor to start-up ownership, and it worked well for generations. It worked so well that it's been replicated in other parts of the world, especially in Asia. Here it's just another part of the local infrastructure that's been disinvested in."

As for local economic development and educational leaders, "they're paralyzed by a lack of good ideas," says Bob. In fact, since 2000 the Western Massachusetts Economic Development Council and its partners, including the University of Massachusetts, have been trying to create a "knowledge corridor" in the region geared to attracting technology-based industries.[2] One of their successes was drawing the supercomputing center to Holyoke (described in the previous chapter). Their commitment to knowledge industries is the flip side of a sweeping rejection of manufacturing more than thirty years in the making. "It's hard for people to see green manufacturing, because the people at the top have said that manufacturing itself is gone, dead," Bob tells me. "They see those big empty factory buildings as giant flashing lights signaling that it's over. But they're wedded to a model of manufacturing that calls to mind big steel and auto assembly plants. They don't seem to understand that manufacturing has changed dramatically—maybe because so many of our [New England] industries, like textiles and paper, died earlier in the twentieth century than they did in the Midwest. Now the work is more dispersed in supply-chain networks and the jobs are more high tech, clean, efficient, and well paid. It's not a scene out of Dickens." His main concern is that the area will lose out on potential growth in manufacturing for lack of skilled workers and that the work that does come into the area won't get to the people who really need the jobs. About the Holyoke computer center, he says, "If they don't build in training linkages to the community of the sort that used to exist as part of the metalworkers' network, no local kid will be able to work in it," and the urban economy will be no better off. While the new imported workforce settles in the suburbs, Holyoke itself will remain mired in poverty and the metro region as a whole won't recover.[3]

We are indeed at a critical crossroads—smaller industrial cities such as Springfield perhaps most crucially of all. After finally accepting that manufacturing is "over," as the United States assumes its rightful place as a knowledge leader in the global economy, American manufacturing could reestablish a foothold in emerging low-carbon industries. And yet that notion bends against received wisdom, long in the making. A 2009 report by MassINC on "building a comprehensive growth strategy" for Springfield captures something of the schizophrenia of our historical moment. While it hammers home the notion that Springfield's future lies with the knowledge economy, it acknowledges that the area's precision manufacturers must find "additional markets," since they haven't been well served by the computer and life science sectors that have been the "focus of statewide economic development efforts for the past two decades." Those manufacturing markets, the report suggests tentatively in a sidebar, could lie with "emerging green technology," since "windmills, mass transit, and a smart energy grid will all require components fabricated by precision manufacturers." But overall, the authors argue forcefully for preparing the workforce, in an equitable manner, for jobs in the new economy dominated by "innovative" knowledge industries.[4]

Springfield, it seems, could point itself in several possible directions. Unlike many smaller industrial cities, mainly in the Midwest, it stopped long ago simply waiting for the manufacturing production cycle to tick upward again. That strategy had plagued these cities from the start, leaving them captive to the fates of one or two large companies buffeted by market forces. Instead, Springfield embraced the knowledge economy and could continue to do so—as urged by the MassINC report—more or less exclusively. But that would ignore the larger historical lesson these places can teach: that economic diversity is crucial to their survival. For reasons that will become clear in this chapter, the idea that cities should cultivate a diversified economic portfolio has become the sole province of high-tech knowledge industries. Oddly, manufacturing, in which smaller industrial cities still have much to offer, is usually left out of the picture. Due to their smaller scale, these cities are also in strong positions to diversify in another way: by supplementing industries that export to the global marketplace with what have been called localist measures that support independently owned retail businesses that generate and retain community wealth. Their smaller urban scale, as we shall see, also puts them in good stead to create excellent, equitable public schools—economic engines in their own right that are essential to stabilizing these places and enriching their civic fabric.

Markets Are Not Free

Although it might seem strange, the best starting point for understanding localist economics lies with the later work of Jane Jacobs—scourge of the "little factory town." It's confusing, too, because her ideas have also been invoked to support agglomerating urban regions in the deregulated global economy, or neoliberalism. Global market restructuring, which began in the 1980s, enhanced the private sector and devolved federal responsibility to state and local governments. Localism, by contrast, is a response to the excesses of globalization. How could both owe an intellectual debt to Jane Jacobs?

With her uncanny powers of observation and synthesis, Jacobs argued in *The Wealth of Cities* (1969) and *Cities and the Wealth of Nations* (1984) that cities rather than nations are the true generators of wealth through their own local economies—their pools of skill, manufactures, materials, and local markets. Writing during a period of urban and industrial decay, Jacobs here took up the age-old question of why some economies grow while others stagnate. The key lies, she argued, not in agricultural surpluses that drive people into the city, as long assumed. Rather, it lies in what Jacobs called "import replacement." It is in thick networks of small firms, diverse in character and concentrated in urban settlements, she argued, that entrepreneurial improvisation, and therefore true economic development, takes place. Crucial here are the "knowledge spillovers" (or "externalities") that ensue among small entrepreneurs working in a variety of fields who find ways of "adding new work to old," thereby spinning off new businesses. Messy, inefficient, and unpredictable, innovation proceeds in spurts, she claimed, in which goods and supplies once imported to the local economy to make goods for export are replaced with locally produced alternatives, with improved design, materials, and production methods as a result of market "feedback loops" that circulate best among many geographically concentrated small firms. Moreover, by bringing wealth back into the community, import replacement has a multiplier effect: it stimulates the growth of small service businesses, from accounting and legal services to retail and restaurants.[5]

Jacobs offered the example of Tokyo's bicycle industry. In the late nineteenth century, Japan began importing bicycles from the West. Repair shops sprang up in Tokyo, at first cannibalizing broken bicycles for parts. When enough of these existed, workshops started producing some of the most commonly used parts locally. More and more parts were made until ultimately Tokyo could produce its own bicycles and export them to other Japanese cities.[6]

Jacobs stresses that import replacement is most likely to take place in large cities, where the greatest number of versatile networks collide, providing conditions for boundless creativity and new production work. Large cities also provide substantial markets, of both consumers and other producers, to absorb new products. For all these reasons, large cities successful at import replacement grow even larger and more creative, not smaller and more specialized, and they are more likely to experiment with solving the problems that all cities face. Import replacement, she argues, therefore "would not be feasible in rural places, company towns or little market towns."[7]

New economy urban theorists wedded to globalization, such as Edward Glaeser and Richard Florida, can find much in Jacobs's work to support their views. They have parlayed her notion that successful urban economies grow only larger—and should—into promoting "star cities" and the megaregion. They have pulled from Jacobs's concept of knowledge spillovers to cultivate business clustering and a global division of labor in which the United States leads in knowledge industries that develop technological efficiencies for the global marketplace. Jacobs's notion of entrepreneurial improvisation and knowledge sharing has shaped Richard Florida's popular concept of the "creative class," along with urban development strategies designed for its sole nurture and delight. Yet what Jacobs loved most about cities—their complexity and variety, their capacity for surprise—has been lost in what has become a fetish for creativity and talent and knowledge, now diminished to align with the imperatives of the new economy and churned out at ever greater levels of efficiency. And it is primed for exhaustion. Where once Florida spoke of the "personality" of cities, he now uses the mechanical language of the engineering lab: "The places that thrive today are those with the highest velocity of ideas, the highest density of talented and creative people, and the highest rate of metabolism."[8]

Indeed, the new economy has reached a point of diminishing returns, and smaller industrial cities are feeling it with full force. It has destroyed American production jobs and put wealth in the pockets of distant global elites, shareholders, and Wall Street overlords who have used it to finance speculative bubbles, creating the temporary illusion of prosperity. It has further consolidated corporate power and destroyed small businesses (even as it has favored small tech start-ups). It has normalized a culture of corruption among political leaders, who have systematically eroded consumer and environmental protections to accommodate market demands. It has reduced politics to economics, economics to growth, and growth to ever greater levels of overfinanced consumption. It has led to obscene levels of wealth inequality. And it has coarsened our civic discourse, offering up a false polarization between "government" and "free markets," as if those were our only choices. Even where trade had been useful between unequal parties, as in the technology gap between Asian countries and the United States, that is no longer the case. As China and India have developed their own high-tech industries with the help of American multinationals, even decently compensated American knowledge jobs are seeing downward wage pressure. The U.S. balance of trade with China rose from about $84 billion in 2000 to almost $227 billion in 2009. In general, Americans must import more and more goods from abroad as its productive capacity has diminished. Yet consumer prices are likely to rise, since Chinese workers in spring 2010 began resisting the low wages and oppressive working conditions that the multinational search for market efficiencies imposes.[9]

With Jacobs's lifelong torrent of words and ideas, it's easy to forget that she was, above all, the sworn enemy of what she called, simply, "sameness," whether it involves stifling innovation in an old company town like Detroit or, I submit, compressing "creativity" into an economic development formula serving the interests of a particular class. Although new economy urbanists can find much in Jacobs to buttress their views, they've lost sight of her warning that ever enlarged economic institutions, gorging on economies of scale, stifle the diversity and complexity so central to her understanding of "how cities actually work." The emerging localist movement seeks to recapture such conditions, while embracing the high value that Jacobs placed on civic improvisation and community self-determination. It seeks to wrest a measure of control from global market forces that favor corporate consolidation, work to homogenize the use of land, and threaten a community's very survival. "An economy that can fill few of the needs of its own people and producers," Jacobs observed, "is not efficient."[10] To redress that state of affairs, localists call for a variation of import replacement—substituting some currently imported goods, sold by identical corporate franchises, with locally produced wares that keep profits circulating at home. They expand creatively, in other words, on Jacobs's idea that "any settlement that becomes good at import replacing *becomes* a city," and continuing to do so *sustains* a city.[11] The Institute for Local Self-Reliance puts the matter succinctly: localism "identifies rules that honor a sense of place and prize rootedness, continuity and stability as well as innovation and enterprise."[12]

Localism is also of particular value to smaller industrial cities, which Florida and others have identified as likely "losers" in the "spatial fix" imposed by today's crisis in global markets, whose great gales of "creative destruction" will, they predict, leave these places in the dust.

Relocalizing the Playing Field

Smaller industrial cities don't have to become ghost towns. If anything, as I have argued, their smaller urban scale, combined with their rich surrounding farmland and manufacturing heritage, could be assets in a low-carbon future. To turn these and other strengths to advantage, the localist movement proposes new rules for retaining the community wealth they generate. It offers what economist David J. Hess calls "alternative pathways" to prosperity that can stabilize these communities, making them more appealing places for export industries to invest in and to carry on the important work of adding new work to old.

As we've already seen, the local food movement has put the idea of keeping agricultural dollars circulating within the regional economy before thousands of urban dwellers who might not otherwise have encountered localism's economic value. The same can be said of the nascent local energy movement, though it has met with less success to date. To these can be added a variety of wealth-retention arrangements, described by David Hess in *Localist Movements in a Global Economy*, that can be useful for cities of all scales. One involves developing business-to-business databases linking local businesses of all sizes to local suppliers. The Oregon Marketplace, for example, preserved as much as $500,000 a year in annual revenue for the state economy. A consortium of some of Chicago's largest corporations set up a similar program for purchasing recycled goods from local suppliers. A more controversial idea is to attach residential restrictions on stock ownership in for-profit corporations, a measure that is permitted under common law and explicitly by many states. The ownership of the Green Bay Packers is structured in this way, which is why the team hasn't become, say, the Minneapolis Packers. Others push for requiring municipalities and locally sited businesses to bank at locally owned financial institutions and credit unions.[13]

These measures reflect a ripening of the localist movement, which has existed in fits and starts since the 1970s. In its earlier days, localism was more self-consciously countercultural, beginning with the off-the-grid communalism inspired by E. F. Schumacher's *Small Is Beautiful* (1973) and the neighborhood-level economic organizing advocated by David J. Morris and Karl Hess in *Neighborhood Power* (1975).[14] Today localism retains some of that anarchistic spirit, but it is less ideological. Rather, it attracts citizens from a spectrum of political positions who acknowledge that exclusive dependence on large businesses successful in the global marketplace has grown too risky: unbridled free markets carry heavy costs, and communities must find ways to hedge themselves against the resulting damage. It prescribes measures that complement the exclusive advantages accorded today's export-oriented, publicly traded corporations. That said, localism is not inconsistent with targeted national tariffs, or federal consumer and environmental regulations, or even national policies that provide incentives for the development of certain industries such as renewable energy. And it positively requires constitutional protections of civil rights, in view of the ugly discriminatory purposes to which local control has been put in the past.

Another way the localist movement has matured—or, rather, expanded to meet today's challenges—is of particular value to smaller cities: its defense of small retail businesses against big-box stores. These category killers are especially pernicious in smaller metro areas. They further dwarf cities of smaller size. Their low prices, made possible by economies of scale and low-wage workers overseas, hold sway over the underemployed and unemployed in older industrial cities, completing an economic loop of impoverishment that makes it hard for opponents to get a hearing. Meanwhile, as we've seen, big-box complexes and the transportation infrastructure they require act as spearheads of sprawl, eating into farmland that could be an important source of prosperity for these local economies and part of their aesthetic appeal.

Not surprisingly, the proliferation of big-box power centers in the 1990s elicited a wail of opposition from small business owners, whose numbers shrank by 100,000 between 1982 and 2002. Outmaneuvered by local chambers of commerce dominated by large retailers and their distant suppliers and service providers, they began organizing independent business councils and launched the first contemporary "buy local" campaigns. Today they are amassed under two umbrella groups: the Business Alliance for Local Living Economies, which is primarily a small-business interest group, and the American Independent Business Alliance, which is more inclined to link the plight of independent business with the environmental degradation and social inequities that big-box economies of scale have left in their wake.[15]

They have much in common, however. The localist independent business movement argues for carving out alternatives to retail consolidation and makes the civic case for place-based economic development policies that give local businesses a fighting chance. First, they seek to level the playing field by countering the tax breaks and subsidies routinely extended to multinational retailers, along with the additional tax burden of providing services to them, with zoning restrictions that limit commercial build-out with big-box chains. Second, they propose plugging the leaky bucket, arguing that large corporate firms leak out community wealth to pay shareholders and distant, handsomely compensated management professionals who don't pay personal taxes in the cities where their firms do business. Locally owned businesses, they claim, plug the leaks. Instead of demanding tax incentives and other subsidies to open shop, small proprietors pay local taxes, provide more and better-paying local jobs, donate more of their profits to local nonprofits, use more local suppliers and services, and have a stake in the aesthetic and environmental quality of their communities—creating an essential feedback loop that they as retail providers share with local consumers.[16]

Stacy Mitchell, author of *Big-Box Swindle: The True Cost of Mega-Retailers and the Fight for America's Independent Businesses*, has been at this fight from her base in Portland, Maine, for a long time. She has conducted countless workshops for the American Independent Business Alliance, which she chairs and that now consists of some 35,000 members organized in 125 local chapters. She also works as a senior researcher for the Institute for Local Self-Reliance's New Rules Project. In an October 2010 telephone interview, I asked Stacy what shape her work takes in smaller cities and whether she thinks urban scale matters at all. "Oh, it matters a great deal," she replied, noting that she was unaware of any studies along these lines and could speak only of her own impressions. "In my experience they are the optimal geography for getting small-business organizing off the ground. There's usually not enough critical mass in small towns, and big cities are really hard because they have to start with a node-based approach, and then each of the neighborhoods has to work with the others more strategically." Besides, she says, "the built environment of small older industrial cities cultivates local relationships, the sort of informal exchange of information, the 'social capital,' that grows a local economy." Sounding much like a new urbanist or smart growth advocate, she emphasizes that small neighborhood businesses also promote walkability and density, reducing carbon-emitting automobile use. She could have pointed to their value in mixed-use development, too, in grounding the commercial side of the "mix."[17]

As Stacy observes in her book, small-to-midsize cities also stand to gain more economically—what she calls the "local premium"—from reining in big-chain retailers than large cities do. "While a big city might house the headquarters of a chain or other large companies, such as advertising firms, that count other global retailers among their major clients," she writes, "this is usually not the case for smaller cities, where most of the dollars flowing into a big-box store are unlikely to ever make their way back into the local economy."[18]

Critics argue that, compared with multinational retailers (with the jobs and commercial tax base they provide), local businesses can contribute to local economies only in atypical communities, such as college towns and bohemian enclaves, where people are already inclined to patronize "indie" enterprises. To put that notion to the test, strategic planning consulting firm Civic Economics conducted a study of metro Grand Rapids, Michigan, to determine the economic impact of currently existing independent pharmacies, grocery stores, full-service restaurants, and banks in the Kent County region. Its results were startling and bore out the findings of similar studies of San Francisco, Austin, and Chicago. The 2008 study showed that if area residents redirected just 10 percent of their spending from chains to locally owned businesses, $137 million in new economic activity would result, including 1,600 new jobs and $53 million in additional payroll. Chain restaurants, for example, return only 37 percent of their revenue to the local economy, compared with independent restaurants, which return 56 percent in wages, locally purchased goods and services, profits, and charitable donations.[19]

This and related studies provide empirical grounding for the otherwise vague sense that big-box stores hollow out local economies. Armed with such findings, the buy-local movement has adopted an incremental, educative, complementary strategy to win political support. With Stacy Mitchell's assistance, localists in Maine, a fairly conservative state, convinced their fellow citizens to pass the country's first state law specifically intended to control mega-retail development. The Informed Growth Act, passed in 2007, requires municipalities to conduct an economic impact study, funded by the developer, for proposed properties larger than 750,000 square feet (about one and a half football fields). The act also requires cities to turn down applications for large-scale-retail land use permits if it can be shown that they would have an "undue adverse impact" on local businesses in the host city and surrounding market area. So far, the act has passed several judicial challenges, which has emboldened other states and municipalities to explore passing similar legislation. (That said, in 2010, the state limited the law's application to new development, not to adaptive reuse, or grayfields, projects.)[20]

In today's neoliberal climate of entrepreneurship and innovation, localist Michael Shuman proclaims, "Let a thousand experiments unfold." More than 500,000 local elected officials govern some 3,000 county governments and about 36,000 municipalities. "Few local governments in the world enjoy the powers that American communities have," he continues. "US mayors and city-council members have a policy tool chest that enables them to invest, contract, zone, tax, lobby, and police. They have the ability to spend public funds on almost anything. While these powers are not unlimited, it's fair to say that the problem facing US local governments is not the absence of powers, but the absence of political will to exercise them."[21]

School Daze

One warm fall day in 2009, I was treated to a lively tour of a midsize Ohio city by a young man on the municipal planning staff. He proudly showed off the metro area's bike paths, the city's impressive mixed-use and affordable neighborhood infill projects, and several community gardens. As we neared the end of our journey, he flashed me a plaintive look and blurted out, "Do you think any of this is really going to make any difference?" Disarmed by his candor, I responded immediately and in kind: "Not unless you can bring back urban retail and good public schools." He agreed.

Millions of Americans who spend their young adult lives in cities and thrive in urban culture would love to remain in cities to raise their children. Many don't because of the poor quality of the public schools. Remarkably, this glaring fact is rarely discussed in the context of urban planning and design. Yet all the walkable neighborhoods, riverfront development, and bike paths in the world will not draw middle-class families to cities without good public schools, which are, among other things, essential engines of economic development.

The troubles besetting the American K–12 educational morass are vexing, longstanding, and speak to the very heart of democratic civic life: What basic knowledge should an educated citizenry share? What are we raising the next generations to do? And how do we do it in a way that is fair and draws out the talents of all children? These are the questions with which local school boards, the most local of our political institutions, must contend. With the 1983 publication of *A Nation at Risk: The Imperative for Educational Reform,* by the National Commission on Excellence in Education, the discretionary power of school boards has been eroded by state and federal curriculum, professional certification, and testing mandates, while public school funding has been fractured by private school voucher programs and charter schools. Yet the taste for these sorts of educational reforms, part of the neoliberal deregulatory ethic, may be subsiding after twenty-five years with little to show for them. A bellwether, perhaps, is the about-face taken by education scholar Diane Ravitch in *The Death and Life of the Great American School System: How Testing and Choice Are Undermining Education.* A former official in George H. W. Bush's Education Department and vocal advocate of No Child Left Behind, she now argues that vouchers, charter schools, teacher merit pay, and escalating testing standards don't work as well as expected—and least of all for the most challenging students for whom these programs are geared. As it turns out, she claims, public schools that maintain high standards and rigorous classroom order, working in tandem with democratically elected teachers' unions, tend to do a much better job overall.[22]

As it happens, school boards in smaller cities hold distinct advantages. The most successful urban public school programs to date, but also by far the hardest to carry out politically, involve creating metropolitan-wide economically integrated school districts. These programs work best in metro areas of smaller scale, such as Raleigh, North Carolina. Raleigh's extraordinary income-based integration program is well known among educators and has been studied from a variety of angles. Yet to date, no one has taken the city's scale into account as a factor in its success. It might be jarring to shift discussion to a city in the South, which lies beyond the geographical scope of this book, but the reason for doing so will become clear.

In *Hope and Despair in the American City: Why There Are No Bad Schools in Raleigh*, education scholar Gerald Grant examines Raleigh's storied achievements in tandem with the unconscionable failures of the Syracuse public school system and links both to the rival fates of their metro economies. Grant, who has spent most of his life in Syracuse and its environs, draws a rich, personally informed historical portrait of his hometown's perilous decline. Let's repeat the drill. Between 1970 and 2000, the Syracuse metro area lost 30 percent of its manufacturing jobs, and the city proper lost almost 40 percent of its population; by 2006, barely half of the city's ninth graders, predominantly poor and minority, graduated from high school. All of this had been decades in the making, beginning with Syracuse's foolhardy decision, replicated in cities across the country in the 1960s, to "clear" the city's historically black neighborhoods to make way for federal urban renewal projects, including a downtown highway. Meanwhile, discriminatory redlining and restrictive racial covenants in the outlying suburbs created the template for the doughnut-like metro geography we know all too well today: city centers filled with dislocated minorities devoid of public resources, surrounded by fluffy white affluent suburbs. Today suburban flight has become a self-perpetuating dynamic driven less by white racism, Grant insists, than by increasingly unruly, low-performing urban public school systems.[23]

Raleigh was headed down the same dreary path, but in 1976 the city and county school boards made the risky decision to merge into one school district under the direction of Wake County, a critical step in averting disaster. This was the South, remember, where school districts had been under judicial pressure since the early 1960s to end legal segregation based on race. The two districts had already achieved considerable racial integration through cross-county busing, made palatable by turning a third of the area's schools into desirable magnet schools, all located along the urban-suburban border, with specialized programs that any parent could select. In the process, the findings of a major 1966 study commissioned by Congress were borne out in Raleigh: what most raised low-income student achievement was not increased funding but attendance at predominantly middle-class schools where students felt safer and gained access to networks of social capital, while more affluent students' achievement remained steady. These findings have been supported time and again over the past forty-five years.[24] The Wake County and Raleigh school districts' decision to merge in 1976 was based in part on a desire to secure these gains, with the recognition that it would be ruinous to the entire regional economy to allow large concentrations of poverty to fester and grow as a result of bad neighborhood schools. The winning rationale that brought Raleigh's business and civic leaders onboard had been articulated in a 1965 Vanderbilt University study concluding that merger not only made good financial sense and would stabilize racial integration, but it also "would be a determining factor in the successful development of the Raleigh Wake County community into a major . . . industrial urban complex."[25]

The full promise of the Wake County–Raleigh merger was not redeemed until 1998, when county superintendent Bill McNeal committed the district to a 95 percent pass rate on state standardized tests—fully four years before the No Child Left Behind law made such matriculation goals mandatory on a national scale. McNeal altered the system in three other ways. First, he made socioeconomic status one of the three main factors in school assignment. He also capped the low-income enrollment for each school at 40 percent, determined by the number of pupils receiving federal subsidies for school lunches. He shook up the teaching culture by giving Raleigh-area teachers more autonomy, while expecting them to work together on teaching strategy, with the most gifted sharing their knowledge with the most challenged. Using computerized testing to identify data-driven weaknesses and individual benchmarks, he rushed extra resources to underperforming students and classrooms. If, after this level of intensive support schools were underperforming, he did not hesitate to shut them down and move students to schools with better track records.

McNeal also staggered the program, implementing it first in the K–8 schools (2000–2003), and then in the high schools, which would thus be unable to blame failure on poor elementary school preparation. The results were impressive, at least for the younger grades. Many schools attained the 95 percent test-pass target, and between 1994 and 2003 third graders' pass figures on math and reading scores rose from 71 percent to 91 percent. For poor children, the figures went from 55 to 80 percent. The same cohort of high school students didn't do nearly as well, but dropout rates for low-income students have steadily declined since 2003. Although there is much room for improvement, Wake has retained good teachers, happy to educate their own kids in the school district, and attracted many high-caliber teachers nationwide.[26]

In the early 1970s, Detroit devised a metropolitan desegregation plan similar to Raleigh's in which the city and its adjoining suburbs would be rendered into pie-shaped school districts. It was challenged, and in 1974, *Milliken* v. *Bradley* went all the way to the Supreme Court, stacked with freshly minted Nixon appointees who had passed the president's antibusing litmus test. The Court ruled the plan unconstitutional since the predominantly white districts did not legally block black families' access to their schools. As a result, we now live with a grand historical irony: public schools in the South are far more integrated than anywhere in the North, whose older industrial cities have become ever more doughnut-like.

When we reflect on the failures of busing, most summon images of the cruel mid-1970s busing wars in Boston, which pitted poor urban communities against one another and asked nothing of affluent suburbanites, whose liberal political leaders supported the plan. Raleigh has demonstrated that there's another way, one that restores the educational center of metropolitan-wide economic development, while ensuring equal access to a good education for all children and maintaining high standards for students, teachers, and administrators alike.

Wake County became the first metropolitan-wide school district to merge voluntarily and to implement a program of income-based balance in 2000. (The first class-based program dates back to 1991 in the blue-collar town of LaCrosse, Wisconsin, where the Laotian Hmong population had grown to 12 percent in a city with a considerable poor-white population.) Others followed, mainly in the South. Chattanooga, Tennessee, a midsize declining industrial city, had tried all the usual stuff—an aquarium on the shore of the Tennessee River, a downtown revitalization initiative—but it wasn't until the city school district merged with the suburbs in 1997 and took seriously the idea that, in the words of the president of the Hamilton County School Board, "We need to be concerned with the overall school system because it's related to our economic health," that the small metro really gained traction. By 2007, the dropout rate had been cut in half, with 75 percent of Chattanooga's students graduating from high school, and the city had been declared a turnaround older industrial city by the Brookings Institution. This is not to say that improving the schools was decisive in the city's current success, but it was one of the critical ingredients.[27]

With the courts ruling against race-based school assignment policies (in cases involving Seattle and Louisville that went to the Supreme Court in 2007), more school districts are looking at socioeconomic integration programs, with strong voluntary and school choice provisions: as of 2010, forty had implemented some version of the model. Among the confusing array of specific policies tailored for local demographics and politics, two considerations stand out. First, some variation of cross-district, urban-suburban cooperation works best as a means of preserving and cultivating an urban middle class, which is essential to the economic health of entire metropolitan regions. And second, the geographical scale of smaller cities makes the necessary student transportation arrangements more feasible:

if students are not assigned to neighborhood schools, they're not likely to travel far from their neighborhoods each day unless they enroll voluntarily in a distant magnet school. In Hartford, Connecticut, a city of some 125,000 with a student poverty rate of about 41 percent, an urban-suburban public school choice program that allows students to move in both directions has been a modest success. Not only do Hartford students have a chance to attend good middle-class suburban schools, but there are long waiting lists of white suburban kids seeking to attend urban magnet schools, four of which were identified as among the best in the country by *U.S. News and World Report*. By contrast, and sadly, many big cities' school districts have such large poor-minority populations that it is extremely difficult to put together the middle-class floor for each school that plans like Wake County's require without unrealistically massive participation by their far-flung suburbs—and ridiculously long bus rides. In New York and Chicago, which have low-income student populations of 74 percent each, "the options have shrunk," observes former Boston school superintendent Tom Payzant. Here, in the critical issue of closing the urban doughnut hole for the sake of the entire economic region, it seems that smaller cities have more options should they choose to exercise them.[28]

Of course, they could choose not to make use of their advantages. In fact, since 2007, in the aftermath of the Supreme Court's rejection of race-based school assignment policies, a number of communities are allowing their schools to resegregate informally by dismantling programs intended to offer children educational pathways out of urban poverty.[29] Nowhere are such emergent path-of-least-resistance policies more disturbing than in Wake County itself, home to one of the best urban school systems in the country. In fall 2009, a slate of candidates dubbed "the Gang of Five" won a majority of seats on the nine-member Wake County school board with the promise of replacing the county's socioeconomic-integration policy with strictly neighborhood-based school assignments. They won 64 percent of the vote with an 11 percent voter turnout; soon after, a school board survey, conducted by the victors, showed that 94.5 percent of Wake County parents were already satisfied with their children's schools. Nonetheless, in March 2010 the Gang of Five began making good on its promise to, in the words of ringleader John Tedesco, "dismantle the social engineering policies of the past under the guise of diversity." A month later, Tedesco gave a speech at a Tax Day Tea Party rally, where he railed against the "social engineers and bureaucrats" who seek to "control the hearts and minds of our children." Tedesco and his fellow candidates received generous direct and indirect support from regional discount retail magnate

Art Pope, who sits on the four-member board of the infamous Koch brothers' Americans for Prosperity—the largest funding source of Tea Party activism and purveyor of the "myth" of global warming. Tedesco and Pope have acknowledged that reversing Wake County's income-integration policy would lead to reconcentrating low-income students in impoverished neighborhood schools. Their proposed remedy for that? To send more government money to city schools—a disingenuous position at odds with Tea Party opposition to government spending.[30]

The conflict over the future of Wake County's public school policies is far from over. By fall 2010, the local NAACP had filed 700 complaints about school reassignments, prompting an investigation by the U.S. Department of Education Office of Civil Rights (ongoing as of April 2011). For now, the drama serves as a reminder of how messy and fractured localism can be, and perhaps most of all in smaller industrial cities. They can be more nimble and their scale makes it possible to try policies that big cities cannot, but their very chances of success can also make them more volatile. Local school boards—the most powerful local political institutions in the land, and probably the most emotionally freighted given their charge over our young—are inherently contentious, to be sure. But what's going on in Wake County also sheds light on how fluid local populations can be, making it ever more difficult to stabilize even highly popular political gains, especially under the tense conditions of what General Electric CEO Jeffrey Immelt calls "the emotional, social, economic reset" brought on by the economic crisis of 2008.

As it turns out, John Tedesco is not local at all. He grew up outside Pittsburgh and spent many years in the greater New York area before moving to Wake County in 2007 to take a job as chief development officer of Big Brothers Big Sisters. He is, in short, a transient professional and something of an educational bureaucrat himself in service to the public-private partnership model.[31] Indeed, Wake County has become a knowledge economy success story, thanks to nearby Research Triangle, as well as to its legendary public school system. Both have attracted a growing number of transient, affluent new economy knowledge workers to the area. It would be all too easy to blame the Gang of Five's win on massive infusions of conservative campaign cash at a time when the GOP's strategic focus was on local races while it's national prospects were dim. A more charitable and compelling explanation has been offered by one of the new school board's most vocal critics. While some voters were driven by "a repugnant ideology," writes Rob Schofield of NC Policy Watch, "many others have gone along with the effort simply because they didn't understand how we got where we are today. These folks don't necessarily share the goals of the conservative ideologues, they're just worried about themselves and their kids and are either too young or too new to the issue (or too new to North Carolina) to understand what's really at stake."[32]

What's at stake, of course, are the impressive gains Wake County made in ending the tragic legacy of racial discrimination. Also at stake are the thoughtful, patient means by which its citizens made it happen: through decades-long community-wide participation, including conservative business leaders and strict old-school teachers, as well as civil rights activists and liberal educators fired with the democratic mission to close the opportunity gap. Something historically novel had taken place here: local cultural memory, long consumed with waving the Confederate bloody shirt, had shifted away from the sectional politics of resentment. It now risks losing those gains to new corrosive forces endangering the area's still fragile more equitable way of life.

"The Curse of Bigness" and the Case for Civic Modesty

Many disparage the Tea Party movement as a revival of populism, by which they mean, vaguely, an unleashing of popular anger, fed by demagogues, directed toward "foreign elements" and "intellectual elites" over whom "the people" feel powerless. That characterization fails to account for the fact that many tea partiers are themselves well-educated "foreigners": affluent white suburbanites, a majority of them men, wedded to the new economy, with its reliance on market solutions for addressing the public interest.[33] Globalization brings with it rootless strangers unfamiliar with the history of their adopted, often transitory new homes. Most are, as Rob Schofield points out, unaware of how and why local arrangements have come about and what would be lost should they be broken. Unleavened by local memory, their politics can harden into the harsh abstractions of ideology, where terms like *liberal* and *government* cannot be unpacked for democratic debate. Nothing could be further from the civic-republican spirit of historic populism, with its wariness of unfettered monopoly and untempered consumer culture—which threaten to dislodge local control and destroy the value of place.

As smaller industrial cities reinvent themselves, they could give way to globalization entirely and let markets sort out the resulting political and cultural realignments. Or they could chart out a third localist way between the welfare state and market triumphalism. Smaller cities, each in its own way, each grounded in its own peculiar history, have a rare historic opportunity to shape a compelling alternative to the cultural juggernaut of globalization, with its self-described global elites who cut deals over lunch in global cities and regard the dumb masses as so much capital fodder for their "smart" stratagems, who in their relentless quest for efficiencies render uniform all ways of life. Free-market extremism has opened up dangerous levels of economic inequality and disemboweled the middle classes.[34] Indeed, cosmopolitan culture itself has been disfigured by this unsustainable race to the bottom line of ever-expanding market returns. It is, in the end, neither liberal nor conservative but, at bottom, aggressively instrumental.

Never before in American history has the instrumentalism of the market gone so unchallenged. The bitter charge of "liberal elitism" voiced by the self-deprecating "rubes in flyover country" is a rhetorical echo of the old populist grievance against eastern industrialists and financiers—the plutocracy, the parasites—who controlled the flow of money from afar and turned Congress into a "millionaires' club." Yet it has become harder than ever to have a serious, long-overdue public debate about the problem of monopoly. Since the 1980s, the cartoonish values imagery of cultural warfare has provided cover for politicians and business leaders eager to unleash the creative destruction of the market—the most unconservative force in history, with profound indifference to values of any kind.[35]

After World War II, the federal government commissioned sociologist C. Wright Mills to write a report on small business and civic welfare. The study was concerned with the substantial gains made by big business during the war. It questioned how such increased levels of economic concentration, already "extremely high," affected the "civic spirit" of "our cities and their inhabitants" of the middle classes. Mills concentrated on three sets of small and medium-size industrial cities, with populations of 25,000 to 50,000, 50,000 to 100,000, and 100,000 to 175,000 in the Northeast and Midwest. Each pair consisted of a big-business city marked by the absentee ownership of a few industrial firms and a small-business city with industrial diversity and a high degree of independent ownership.

Smaller industrial cities dependent on big business, Mills concluded, suffered from the rule of distant corporate elites who found salary holders to do their bidding in local politics and chambers of commerce, introduced chaos with factory shutdowns, and offered little patronage to local cultural institutions. Small-business cities, Mills found, were far more likely to have a thriving democratic culture. Turning the old sneer about Babbittry on its head, he argued that the success of the independent "business enterpriser," in contrast with the "business careerist," is "locally rooted and locally oriented," and thus "his own business success is linked to his participation in civic affairs." Competition in "civic activity" that improves one's "own economic and position" and "social prestige," Mills argued, is good for democracy. Likewise, Mills viewed the "wives of local businessmen" (during this lull in modern feminist revolt) as indispensable contributors to the civic welfare. Their "rivalry" and quest for status cultivated "leadership in civic enterprises" in "education, health, and charities," and in building libraries and local arts institutions. Most striking, perhaps, for contemporary readers who bemoan the brain drain in today's small industrial cities, Mills pointed out that small-business cities were more likely to retain their "bright young men," whereas those dependent on big business tended to put their children on a corporate path from which they were less likely to return; if they did, they were more tied to their corporate identity than to their civic one and more likely to live in the suburbs.[36]

The key terms here are *industrial diversity* and *civic spirit*. Mills worked within a then-dying liberal tradition that had deep roots in antimonopoly thought, an heir to the Jeffersonian ideal that prized decentralized power and appealed to the working people, small business owners, farmers, and shopkeepers most endangered by the "great trusts." Conservative in its instincts—aware of what was being lost to the inexorable progress of the market if left unfettered—this earlier version of populist progressivism began splitting off in an array of reactionary directions in the 1920s: in the nationalist America First movement in foreign policy, in opposition to labor unions' bargaining rights, and ultimately in the states' rights rhetoric opposed to federal protections of African Americans' civil rights. It would fall to the 1960s New Left, which coined the term *liberal elitism* and was inspired by C. Wright Mills's studies of the power elite, to revive antimonopoly politics and, in the 1970s, to mount a decentralized localist movement in response to the "ecological" crisis unleashed by unrestrained corporate-industrial power.

Today's localist movement is part of that longer tradition, which is little understood and well worth recovering, for it could appeal to both liberals and conservatives during our own era of economic upheaval and political crisis. It is exemplified by insurgent Republican Robert M. "Fighting Bob" La Follette Sr., the great progressive Wisconsin governor, U.S. representative, and senator who, between 1885 and 1925, railed against the "vast corporate combinations" while introducing to his state reforms intended to balance their power: a graduated income tax, direct democracy, a comprehensive civil service system, and workers' compensation. La Follette also supported women's suffrage and federal child labor laws, and was rare among his contemporaries in pushing for Indian land rights and black civil rights—even though Wisconsin itself was extremely white and he had nothing to gain politically in doing so. Ever independent minded, La Follette managed to avoid the invasive moralism so common among progressives—and commonly resented—as reflected in the debacle of Prohibition. La Follette focused instead on what he believed were weightier public issues that secured economic fair play and civil liberties. To that end, he championed the anti–chain store movement of the 1920s and 1930s—precursor to today's buy-local campaigns—which challenged the wave of retail consolidation made possible by truck transport and new credit arrangements for bulk purchasing: by 1930, A&P alone had 15,700 stores commanding 40 percent of grocery market share and was the fifth-largest industrial corporation in the country. The movement met with some success in nongrocery retail for a time, but obviously it lost the longer battle.[37]

Another titan of the antimonopoly tradition, Louisville-bred "people's lawyer" and later Supreme Court justice Louis D. Brandeis, penned the painstakingly researched economic populist tract *Other People's Money and How the Bankers Use It*, a book still well worth reading today for its probing analysis and moral condemnation of financial oligarchy. Brandeis too was a vigorous judicial supporter of the anti–chain store movement. A close adviser to President Woodrow Wilson, he brokered the Federal Reserve Act, creating a decentralized counterbalance to the money trust. And the reissue of *Other People's Money* in 1933 tipped political debate in favor of the Glass-Steagall Act, which separated commercial and investment banking operations (repealed in 1999, leading to the 2008 meltdown of financial markets) and established the Federal Deposit Insurance Corporation for commercial banks. Considered one of three liberals on the federal bench, Brandeis was a fierce advocate of free speech and privacy protections and was especially critical of the New Deal from the bench, favoring state jurisdiction over excessive federal centralization. Brandeis was not uniformly opposed to monopoly and certainly recognized the value of economies of scale in some industries. As a matter of political philosophy, however, he feared that "the curse of bigness" would overwhelm democratic civic spirit, turning Americans into passive consumers of both politics and culture.[38]

Lewis Mumford, culture critic and urbanist, is not usually linked with these lions of antimonopoly progressive populism. Yet his ideal of spatial democracy, providing for ecological regionalism and urban decentrism, also belongs to this tradition. Mumford's critique of the giantism and cultural uniformity of the sprawling metropolis is of a piece with Brandeis and La Follette's opposition to unbridled monopoly. Indeed, he argued, the modern metropolis is the urban form of concentrated finance monopoly, which extended its tentacles beyond the city to control independent businesses throughout the land. "To complete the process of metropolitan monopoly," he wrote, "its one-sided control must be pushed even further: by buying up and assembling local enterprises, forming chains of hotels or department stores that may be placed under centralized control and milked for monopoly profits."[39] Mumford's plea for limits to urban growth, for a decentralized mosaic of smaller cities, is at bottom a call for recovering local civic participation from the thrall of financial consolidation and control.

Market efficiency and the drive toward growth has and always will exist in tension, shifting over time, with the civic and spiritual principles through which communities make sense of the world. The decentralists of the populist and progressive movements were unable to rein in the trusts, but they did succeed in what historian Richard Hofstadter called "the important intangibles of political tone." The popular "mood" of the era, he argued, threw financial interests "intermittently on the defensive."[40] Today, in Hofstadter's turn of phrase, we need to set a political tone in which sheer economic might does not reign supreme to the exclusion of all other values. As we've seen, today's localists call for new rules that even the playing field and make it possible for small, independent businesses to function again in communities beloved by their citizens, for whom stewardship of place is a moral obligation of the first order. In that world, businesses dealing in global markets are more than welcome, but not with excessive subsidies and exorbitant government handouts that give them an unfair advantage.

The liberal populist-progressive tradition of decentralization, with its conservative instincts of independence, preservation, and fair play, offers a body of thought for imagining a bright, more diversified future for smaller industrial cities. By developing both knowledge- and manufacturing-based low-carbon industries, relocalizing agriculture and food systems, developing appropriate transportation systems, reviving local retail and curbing sprawl, and putting their smaller scale to advantage in creating truly great public schools, these places could thrive economically, with productive work that people of all classes can do with integrity. In the process, they could also thrive culturally, as places that testify to the idea that a sustainable economic culture allows us to flourish as a people, in all our manifold ways. Smaller urban scale can be a strength in a truly democratic, environmentally sustainable national culture—not in competition with global cities but with a fair claim to respect in the eyes of the world.

Notes

Introduction

1. James Howard Kunstler, *The Geography of Nowhere: The Rise and Decline of America's Man-Made Landscape* (New York: Simon & Schuster, 1993). Kunstler is among the few analysts who view smaller cities as having notable advantages over larger ones in a low-carbon future. See especially *The Long Emergency: Surviving the Converging Catastrophes of the 21st Century* (New York: Atlantic Monthly Press, 2005), 250–256 and KunstlerCast #15: "Reactivating Small Cities," May 21, 2008.

2. There is no consensus among demographers and other statistical experts about just what constitutes a small or midsize city. Both the U.S. Census and the United Nations Population Division put the floor for large cities at 500,000. Above that number, many current students of urbanism make further distinctions among large cities and megacities, or global cities, the latter of which relate more to one another in the world economic system than they do to their host countries or states. There is much debate, however, about where to draw the line between a large town and a small city or, for that matter, between a small city and a midsize one. The U.S. Department of Housing and Urban Development regards small cities as any settlement below 50,000 and leaves it to the states to determine what a "town" is. Complicating matters, the states use numerous figures for drawing that line. In Ohio, a city must consist of at least 5,000 people, whereas in Nebraska, that number is only 800. Also, many argue that given the fact of suburban sprawl, the urban boundary itself is irrelevant to the task of defining size and that what should be considered is either density or the size of the metropolitan area. Since 2003, the U.S. Census and Office of Management and Budget have made a distinction between metropolitan and micropolitan statistical areas. The definition of a micropolitan area is a settlement organized around a core of between 10,000 and 49,999 people. Given all the confusion and debate, I'm using the latter figure to distinguish between a large town and a small city, and I'm using the United Nations and the U.S. Census Bureau to delimit the size of the places I examine.

My own working definition of smaller cities, by which I mean small-to-midsize cities, is based on several considerations. The first is historical: the cities I'm looking at, once grand, lost population to deindustrialization and the rise of the Rust Belt, yet they have the infrastructure capacity to grow again or shrink to a sustainable size. For that reason, my definition is scaled over time. Buffalo, for example, was considered a big city in 1950, with a population of 580,000; today that number is estimated to be around 270,000, clearly a midsize city. For the same reason, it doesn't make a whole lot of sense to use density as a measure. Many of these settlements have lost density, and yet they are cities. Finally, while identifying statistical areas might make sense for federal agencies doling out transportation dollars and other types of funding, such designations make no distinction between cities with their suburbs—a distinction that I think is crucial to an understanding of both forms, which is further discussed in terms of new urbanist transect theory in chapter 3.

In any case, almost without exception, every study of small cities begins with a ritual lunge at definition. For a good general survey of the subject's slipperiness, see John D. Buenker and Theodore Mesmer, "A Separate Universe? An Exploratory Effort at Defining the Small City," *Indiana Magazine of History* 99:4 (2003): 331–352.

3. The literature on the dangers of oil dependency is vast and many years in the making. Work that shaped my own views includes William Behrens III, Donnella Meadows, Dennis Meadows, and Jørgen Randers, *Limits to Growth: A Report for the Club of Rome's Project on the Predicament of Mankind* (New York: Universe Books, 1972); Bill McKibben, *The End of Nature*, 2nd ed. (New York: Anchor Books, 1999); Elizabeth Kolbert, *Field Notes from a Catastrophe: Man, Nature, and Climate Change* (New York: Bloomsbury Press, 2006). James Howard Kunstler, "The Long Emergency," *Rolling Stone*, March 23, 2005; Michael Pollan, The *Omnivore's Dilemma: A Natural History of Four Meals* (New York: Penguin, 2006); Jay Inslee and Bracken Hendricks, *Apollo's Fire: Igniting America's Clean Energy Economy* (Washington, DC: Island Press, 2008). Thomas Friedman, *Hot, Flat, and Crowded: Why We Need a Green Revolution and How It Can Renew America* (New York: Farrar, Straus and Giroux, 2008) does a good job of emphasizing oil dependency as a national security threat. For a good conservative appraisal of the status of world oil supplies and the demands of energy security, one that doesn't simply deny the finitude of oil, see Daniel Yergin, "Ensuring Energy Security: Old Questions, New Answers," *Foreign Affairs* 85, no. 2 (2006): 69–82. See also Russell Gold and Ann Davis, "Oil Officials See Limit Looming on Production," *Wall Street Journal*, November 19, 2007, A-1.

4. These figures, which have been quoted widely, originated with the Clinton Climate Initiative upon launching the Large Cities Climate Leadership Group (or C40) in 2006. They have been challenged, notably by David Satterthwaite and David Dodman, who argue that the urban greenhouse gas emissions figure is closer to 40 percent. See Satterthwaite and Dodman, "Are Cities Really to Blame?" *Urban World* (March 2009): 12–13.

5. Richard Morin and Paul Taylor, *Suburbs Not Most Popular, But Suburbanites Most Content* (Washington, DC: Pew Research Center, 2009). See also Reid Ewing, "Is Los Angeles–Style Sprawl Desirable? *Journal of the American Planning Association* 63, no. 1 (1997): 101–126.

6. "Why isn't there more new urbanism?" thread, Gristmill: The environmental news blog, post by jrslide, 8:43 a.m., March 21, 2006.

7. See, for example, Herrington J. Bryce (ed.), *Small Cities in Transition: The Dynamics of Growth and Decline* (Cambridge, MA: Ballinger, 1977); Herrington J. Bryce, *Planning Smaller Cities* (Lexington, MA: D. C. Heath, 1979); Brian Berry, *Geography of Market Centers and Retail Distribution* (Englewood Cliffs, NJ: Prentice Hall, 1967); James Vance, *The Merchant's World: The Geography of Wholesaling* (Englewood Cliffs, NJ: Prentice Hall, 1970); Mary Procter and Bill Matuszeski, *Gritty Cities* (Philadelphia: Temple University Press, 1978).

8. See Beth Siegel and Andy Waxman, *Third-Tier Cities: Adjusting to the New Economy*, U.S. Economic Development Administration: Reviews of Economic Development Literature and Practice, No. 6 (2001), whose bibliography includes a sprinkling of references to studies of small cities in the 1990s; M. Burayidi (ed.), *Downtowns: Revitalizing the Centers of Small Urban Communities* (New York: Routledge, 2001); George A. Erickcek and Hannah McKinney, "'Small Cities Blues': Looking for Growth Factors in Small and Medium-Sized Cities" (working paper 04-100, Upjohn Institute for Employment Research, Kalamazoo, MI, June 2004); Lorlene Hoyt and André Leroux, *Voices from Forgotten Cities: Innovative Revitalization Coalitions in America's Older Small Cities* (New York: PolicyLink, 2007); Radhika Fox and Miriam Axel-Lute, *To Be Strong Again: Renewing the Promise in Smaller Industrial Cities* (New York: PolicyLink, 2008); David Bell and Mark Jayne (eds.), *Small Cities: Urban Experience Beyond the Metropolis* (London: Routledge, 2006), and "Small Cities? Towards a Research Agenda," *International Journal of Urban and Regional Research* 33 (2009): 683–699; André Leroux, "New England's Small Cities: A Mostly Untapped Resource" (Boston: Federal Reserve Bank of Boston, 2010); Benjamin Ofori-Amoah (ed.), *Beyond the Metropolis: Urban Geography as If Small Cities Mattered* (Lanham, MD: University Press of America, 2007); William Garett-Petts (ed.), *The Small Cities Book: On the Cultural Future of Small Cities* (Vancouver: New Star Books, 2005); Timothy Mahoney, "The Small City in American History," *Indiana Magazine of History* 99 (2003): 311–330; James J. Connolly, "Decentering Urban History: Peripheral Cities in the Modern World," *Journal of Urban History* 35 (2008): 3–14. Jeffrey Bean, "Holistic Revitalization in Small Post-Industrial Cities: Tools for Urban Housing Development," (master's thesis, MIT, Department of Urban Studies and Planning, 2009).

9. Louis Wirth, "Urbanism as a Way of Life," *American Journal of Sociology* 44:1 (July 1938): 1–24.

10. Connolly, "Decentering Urban History," 4. All of the twentieth-century social science disciplines were profoundly shaped by sociologist Ferdinand Tönnies's theory, advanced in 1887, that the arc of modernity can be traced through the shift from small face-to-face communities with common mores (*gemeinschaft*) to impersonal urban-industrial society (*gesellschaft*).

11. Jane Jacobs, *The Death and Life of Great American Cities* (New York: Random House, 1961); Richard Florida, *The Rise of the Creative Class and How It's Transforming Work, Leisure, Community, and Everyday Life* (New York: Basic Books, 2002), and *The Great Reset: How New Ways of Living and Working*

Drive Post-Crash Prosperity (New York: HarperCollins, 2010), which expanded on "How the Crash Will Reshape America," *Atlantic Monthly* (March 2009): 23–36. Florida began writing about megaregions in the lead-up to *Who's Your City? How the Creative Economy Is Making Where You Live the Most Important Decision of Your Life* (New York: Basic Books, 2008). Edward L. Glaeser, "Why Has Globalization Led to Bigger Cities?" *New York Times* Economix blog, May 19, 2009; Joel Kotkin, *The New Geography: How the Digital Revolution Is Reshaping the American Landscape* (New York: Random House, 2001), and *The Next Hundred Million: America in 2050* (New York: Penguin, 2010). Most of Kotkin's views on the suburbs appear in his many *Forbes* magazine columns and newspaper articles, as well as on his blog, newgeography.com. His views are well summarized, and challenged, by Yonah Freemark, "Why Is Joel Kotkin Extolling the Virtues of Suburbia?" *Next American City* blog, March 2, 2010, http://americancity.org/columns/entry/2092/ (accessed May 12, 2010). Joel Garreau, *Edge City: Life on the New Frontier* (New York: Anchor Books, 1991); Alan Ehrenhalt, "Trading Places: The Demographic Inversion of the American City," *New Republic*, August 13, 2008. An excellent practitioner's guide to smart growth in rural areas, published in 2010, makes useful distinctions between small towns and small cities, reflecting a possible shift toward recognizing the distinct attributes (and challenges) of small and midsize cities among sustainability advocates. See Nadeja Mishkovsky, Mathew Dalbey, Stephanie Bertaina, Anna Read, and Tad McGalliard, *Putting Smart Growth to Work in Rural Communities* (Washington, DC: International City/County Management Association, 2010).

12. Nancy Oakley, "Keys to the City of Asheville," *USAirways Magazine* (September 2009): 52–66.

13. See Buenker and Mesmer, "A Separate Universe?" and Mahoney, "The Small City in American History."

14. On the extent of eastern investment in and management of the railroad system that, in effect, created smaller heartland industrial cities and integrated them into a national market system, see William Cronon, *Nature's Metropolis: Chicago and the Great West* (New York: Norton, 1991), 82–83; Jon C. Teaford, *Cities of the Heartland: The Rise and Fall of the Industrial Midwest* (Bloomington: University of Indiana Press, 1993), 33–40.

15. Quoted in Emily Talen, *New Urbanism and American Planning: The Conflict of Cultures* (New York: Routledge, 2005): 116 (see also 77, 67–77, 111–143, 195–198); William H. Wilson, *The City Beautiful Movement* (Baltimore, MD: Johns Hopkins University Press, 1989). For the fullest contemporary statement of City Beautiful ideas, see Charles Mulford Robinson, *The Improvement of Towns and Cities; or, The Practical Basis of Civic Aesthetics* (G. P. Putnam's Sons, 1903). On the lingering presence of historic building stock in the core of smaller cities, see Kent Robertson, "Development Principles for Small Cities," in Michael A. Burayidi, (ed), *Downtowns: Revitalizing the Centers of Small Urban Communities* (New York: Routledge, 2001), 9–22.

16. John Nolen, *Replanning Small Cities: Six Typical Studies* (New York: B. W. Huebsch, 1912), 2.

17. Ibid., 5, 154–155. Though mostly redundant, see also *New Ideals in the Planning of Cities, Towns, and Villages* (New York: American City Bureau, 1919).

18. For a full list of Nolen's commissions, see the appendix to John Nolen, *New Towns for Old: Achievements in Civic Improvement in Some American Small Towns and Neighborhoods*. ed. and introduced by Charles D. Warren. (Amherst: University of Massachusetts Press, 2005).

19. Roderick D. McKenzie, *The Metropolitan Community* (New York: McGraw-Hill, 1933). See also Mel Scott, *American City Planning since 1890* (Berkeley: University of California Press, 1971). Bernard J. Newman, review of Russell Van Nest Black, "Planning for the Small American City," *Annals of the American Academy of Political and Social Science* 174, no. 1 (1934): 194.

20. Park Dixon Goist, *From Main Street to State Street: Town, City, and Community in America* (Port Washington, New York: Kennikat Press, 1977), 3–58. Although Goist does a good job documenting these fine but critical distinctions, he doesn't probe the small town–big city polarity since his interest lies in the effect of modern industry on the search for community.

21. Sinclair Lewis, *Babbitt* (New York: Harcourt Brace Jovanovich, 1922), 311. For a discussion of Lewis's plans for *Babbitt*, see Richard Lingeman, *Sinclair Lewis* (New York: Random House, 2002), 171–174. Lewis's other major biographer captures the way contemporary readers grasped the importance of *Babbitt*'s setting but doesn't assign much importance to it. See Mark Schorer, *Sinclair Lewis: An American Life* (New York: McGraw-Hill, 1961), 343–363. See also James M. Hutchisson, *The Rise of Sinclair Lewis, 1920–1930* (University Park: Penn State University Press, 1996), whose exhaustive research shows that Lewis thought of *Babbitt* "as much about the phenomenon of the American city as about 'The Tired Businessman'" (57) and that he "planned to feature Zenith more prominently in the novel than he ultimately did" (66) He also intended originally to include an appendix titled "Main Street vs. the Boulevard vs. Fifth Ave.," Hutchisson reports (57). Richard Lingeman, *Small Town America: A Narrative History, 1620–The Present* (Boston: Houghton Mifflin, 1980); Henry F. May, *The End of American Innocence: A Study of the First Years of Our Own Time, 1912–1917* (New York: Knopf, 1959).

22. Robert S. Lynd and Helen Merrell Lynd, *Middletown: A Study in Modern American Culture* (New York: Harcourt, Brace, and World, 1929), and *Middletown in Transition: A Study in Cultural Conflicts* (New York: Harcourt, Brace, 1937). See also Richard Wightman Fox, "Epitaph for Middletown: Robert S. Lynd and the Analysis of Consumer Culture," in Richard Wightman Fox and T. J. Jackson Lears (eds.), *The Culture of Consumption: Critical Essays in American History, 1880–1980* (New York: Pantheon, 1983).

23. See C. Wright Mills, *White Collar: The American Middle Classes* (New York: Oxford University Press, 2002), 250–251, and *The Power Elite*, 30–46, 386–388.

24. See Walter Christaller, *Central Places in Southern Germany*, trans. Carlisle W. Baksin (Englewood Cliffs, NJ: Prentice Hall, 1966); Paul M. Hohenberg and Lynn Hollen Lees, *The Making of Modern Europe, 1000–1950* (Cambridge, MA: Harvard University Press, 1985); and Brian J. L. Berry and Chauncy D. Harris,

"Walter Christaller: An Appreciation," *Geographical Review* 60, no. 1 (1970): 116–119.

25. Raymond Mohl and Arnold Richard Hirsch (eds.), *Urban Policy in Twentieth-Century America* (New Brunswick, NJ: Rutgers University Press, 1993); Robert Wood, "Cities in Trouble," *Domestic Affairs* 1 (Summer 1991): 223–235; Dolores Hayden, *Building Suburbia: Green Fields and Urban Growth, 1820–2000* (New York: Vintage, 2003), 128–180. For a decent overview of the clash between proponents of urban growth boundaries and property rights activists, see Anthony Flint, *This Land: The Battle over Sprawl and the Future of America* (Baltimore, MD: Johns Hopkins University Press, 2006).

26. See Hoyt and Leroux, *Voices from Forgotten Cities.*

27. Aaron Renn, author of The Urbanophile blog, e-mail exchange with the author, March 3, 2009.

28. U.N. Department of Public Information, press conference on 2009 Revision of World Urbanization Prospects, March 25, 2010.

29. Christina Larson, "China's Grand Plans for Eco-Cities Now Lie Abandoned," *Yale Environment 360*, April 6, 2009, http://featured.matternetwork.com/2009/4/chinas-grand-plans-eco-cities.cfm (accessed September 30, 2010). For a good example of the breathless giddiness with which Western planners greeted news of China's urbanization plan, see the Dynamic City Foundation, established in 2003 to launch what it branded China 2020: http://www.dynamiccity.org/summary.php (accessed September 30, 2010).

30. See "China Builds Model Low-Carbon City in Xinjian's Turpan," *China Daily*, May 12, 2010, http://www.chinadaily.com.cn/business/2010-05/12/content_9841410.htm (accessed September 30, 2010); World Bank, "Transport in East Asia and the Pacific: China Overview" (Washington, DC: World Bank, 2010); Keith Bradsher, "China Sees Growth Engine in a Web of Fast Trains," *New York Times*, February 12, 2010.

31. Sustainable Lawrence, "The Eco-Municipality Model of Sustainable Community Change" (August 2, 2006); American Planning Association, "Policy Guide on Planning for Sustainability" (April 17, 2000). See also the Institute for Eco-Municipalities home page at http://www.instituteforecomunicipalities.org/Home_Page.html; Heike Mayer and Paul Knox, *Small Town Sustainability: Economic, Social, and Environmental Innovation* (Basel, Switzerland: Birkhäuser, 2009); and Timothy Beatley, *Green Urbanism: Learning from European Cities* (Washington, DC: Island Press, 2000).

32. The Obama administration has also taken a "place-based" approach—in contrast with an individual-project-based approach—to funding interagency urban development programs. While its work is piecemeal in the absence of congressional low-carbon policy commitments, the Partnership for Sustainable Communities has been particularly successful in working with formerly neglected community development leaders to target funding for HUD comprehensive planning challenge grants, EPA brownfields redevelopment work, and DOT TIGER II stimulus grants, most of which went to transportation infrastructure repair and urban public transit. See HUD-DOT-EPA Partnership for Sustainable Communities at http://epa

.gov/smartgrowth/partnership/ and Willy Staley, "Urban Nation," *Next American City* 29 (Winter 2010): 7–12.

Chapter 1

1. The term "shapeless giantism" is Lewis Mumford's. See *The Culture of Cities* (New York: Harcourt, 1938), 233. Note that in *The City in History*, a much revised version of the earlier work published in 1961, he uses the term "sprawling giantism" (543). Interview with Julie Backenkeller, November 12, 2009.

2. http://www.1kfriends.org/publications/newsletters/eco-municipalities-sustainable-community-development/ (accessed March 20, 2011). Since 2006, when 1000 Friends of Wisconsin created its eco-municipality program, it has launched a similar Green Tier Communities program in formal partnership with the state Department of Natural Resources, which is also described on the organization's website.

3. Marcia Nelesen, "State Program Could Help Save Farmland," *GazetteExtra.com* , February 16, 2010, http://gazettextra.com/news/2010/feb/16/state-program-could-help-save-farmland/ (accessed October 4, 2010).

4. See Emily Talen, *New Urbanism and American Planning: The Conflict of Cultures* (New York: Routledge, 2005), 127, 138–139. Talen coins the term "doctrine of appropriateness" to describe a strain of thinking that blended concerns with both beauty and efficiency in urban design, and views John Nolen, Raymond Unwin, Charles Mulford Robinson, and Daniel Burnham as its principal practitioners.

5. "City of Janesville, Wisconsin, Comprehensive Financial Report for the Year Ending December 31, 2008," 7.

6. Interview with Brad Cantrell, November 12, 2009. A Standard & Poor's November 4, 2009, credit report gives the city an AA bond rating, in part based on jobs available in Rockford and Madison.

7. "City of Janesville Comprehensive Plan" (2009), http://www.ci.janesville.wi.us/citysite/DeptHome.aspx?Dept=ComprehensivePlan (accessed February 2, 2009). The best source on the history of Wisconsin's 1999 smart growth legislation is the website for 1000 Friends of Wisconsin, which had a leading hand in shaping and passing the bill: http://www.1kfriends.org/what-we-do/community-planning/ (accessed March 20, 2011). For a more comprehensive characterization of smart growth, see Dan Emerine, Christine Shenot, Mary Kay Bailey, Lee Sobel, and Megan Susman, *This Is Smart Growth* (Washington, DC: Smart Growth Network, 2006), and Andres Duany, Jeff Speck, and Mike Lydon, *The Smart Growth Manual* (New York: McGraw-Hill, 2009).

8. E-mail from Julie Backenkeller, November 8, 2009. See also Sustainability Committee member Alex Cunningham's "Preliminary Comments on the Draft Documents: Existing Conditions Report (June 8, 2007) and the Janesville Comprehensive Plan (August 2008)," revised November 3, 2008, which Julie Backenkeller made available to me.

9. US 14/Wis 11 Corridor Advisory Committee Meeting #5 minutes, August 21, 2008.

10. Richard C. Longworth, *Caught in the Middle: America's Heartland in the Age of Globalism* (New York: Bloomsbury, 2008), 161.

11. For a good overview of state annexation policies, which vary widely among states and regions, see David Rusk, *Annexation and the Fiscal Fate of Cities* (Washington, DC: Brookings Institution, 2006).

12. Interview with Mike Saunders, November 13, 2009.

13. "Draft of Town of LaPrairie's Comprehensive Plan, Chapter 3: Agricultural, Natural and Cultural Resources," (2009), 3.

14. Nelesen, "State Program Could Help Save Farmland."

15. Joel Kotkin, *The City: A Global History* (New York: Random House, 2005), 85–96.

16. Ibid., 86.

17. Lewis Mumford, "Home Remedies for Urban Cancer," in Lewis Mumford, *The Urban Prospect* (London: Secker & Waburg, 1968), 187.

18. Jane Jacobs, *The Death and Life of Great American Cities* (New York: Vintage, 1961), 7–25, 218–221.

19. The term *growth machine* comes from Harvey Molocht and John Logan, *Urban Fortunes: The Political Economy of Place* (Berkeley: University of California Press, 1980).

20. Jacobs, *Death and Life*, 20.

21. See Ebenezer Howard, *Garden Cities of Tomorrow* (London: Faber and Faber, 1902), a revised and more widely distributed version of *To-morrow: A Peaceful Path to Real Reform*.

22. Jacobs, *Death and Life*, 25. For an excellent, succinct analysis of Jacobs's distorting conflation of all types of planning, see Mark Luccarelli, *Lewis Mumford and the Ecological Region: The Politics of Planning* (New York: Guilford Press, 1995), 202–208. For the best—and most sympathetic—discussion of Howard and the garden city movement, see Peter Hall, *Cities of Tomorrow*, 3rd ed. (Victoria: Blackwell, 2002), 88–141.

23. Lewis Mumford, "The City," in Lewis Mumford, *City Development: Studies in Disintegration and Renewal* (New York: Harcourt, Brace, 1945), 3–25.

24. See, for example, Andrew A. Meyers, "Invisible Cities: Lewis Mumford, Thomas Adams, and the Invention of the Regional City, 1923–29," *Business and Economic History* 27 (1998): 292–306.

25. Writing in the 1970s, historian Carl Sussman was one of few students of Mumford who recognized how central "interconnected small cities" were to Mumford's urbanism. "If suburbanization," he writes, "has preempted Mumford's label for their regional alternative—the fourth migration—then perhaps the RPAA's vision can someday be realized as the fifth migration." See Carl Sussman (ed.), *Planning for the Fourth Migration: The Neglected Vision of the Regional Planning Association of America* (Cambridge, MA: MIT Press, 1976), 4, 44.

26. Mumford, "The Social Foundations of Post-War Building (1942)," in *City Development*, 186.

27. Mumford, "The City," 19.

28. Mumford, *Culture of Cities*, 254.

29. Mumford, "A Brief History of Urban Frustration (1967)," in *The Urban Prospect*, 219.

30. Mumford, *Culture of Cities*, 326–327.

31. Mumford, "The Highway and the City," *Architectural Record* (1958); reprinted in *Urban Prospect*, 92–107.

32. On Mumford's later emphasis on recentralizing existing exurban settlements and his appreciation of Princeton, New Jersey (for example), as already approximating "new town" garden city principles, see Luccarelli, *Lewis Mumford and the Ecological Region,* 209–214. Mumford did not involve himself in either the short-lived Great Society "new town" federal loan program or the Reston, Virginia, and Columbia, Maryland, "garden city" projects. His biographer, Donald Miller, speculates that he was not brought into the projects early enough to be interested, but the broader evidence suggests that his interest had turned from planning cities wholesale to modifying existing settlements through incremental change. See Miller, *Lewis Mumford: A Life* (New York: Grove Press, 2002), 488.

33. Mumford, "The Theory and Practice of Regionalism," *Sociological Review* 20 (January 1928): 25–26.

34. Mumford, *Culture of Cities*, 397.

35. Ibid., 485–491.

36. For more a more detailed treatment of these failures, see Talen, *New Urbanism and American Planning*, 225–229.

37. Mumford, "A Brief History of Urban Frustration," in *Urban Prospect*, 217.

38. Ibid., 216. Several excellent studies of ecological regionalism and green cities focus on Howard, Mumford, and Geddes much more extensively than I have here. Curiously, they apply their contemporary analysis to either large cities or urban settlements in general, regardless of scale. The small city's distinct attributes, so central to these thinkers, is obscured. See, for example, Kermit C. Parsons and David Schuyler (eds.), *From Garden City to Green City: The Legacy of Ebenezer Howard* (Baltimore, MD: Johns Hopkins University Press, 2002); Luccarelli, *Lewis Mumford and the Ecological Region*; Robert Fishman, "The American Garden City: Still Relevant?" in Stephen V. Ward (ed.), *The Garden City: Past, Present, and Future* (London: Chapman & Hall, 1992), 146–164; and Ben Minteer, *The Landscape of Reform: Civic Pragmatism and Environmental Thought in America* (Cambridge, MA: MIT Press, 2006).

39. Institute of Transportation Studies (Berkeley) and ICF Consulting, "Metropolitan-Level Transportation Funding Sources" (American Association of State Highway and Transportation Officials, December 2005), 9. See also Mark Solof, History of Metropolitan Planning Organizations (Newark: New Jersey Transportation Planning Authority, 1998); Texas Transportation Institute,

"Metropolitan Transportation Planning Issues: A Primer/Anthology of Small and Medium MPOs" (Transportation Model Improvement Program: April 1999).

Chapter 2

1. Interview with Hunter Morrison, October 22, 2009. As indicated, some of the following quotes come from a follow-up telephone interview on October 5, 2010.

2. For a typical comparison of high-speed rail to the interstate highway system, see this blog post by Transportation Secretary Ray LaHood: "High-Speed Rail; No Turning Back on American Jobs, Economic Opportunities, Mobility Gains," October 5. 2010, http://fastlane.dot.gov/2010/10/high-speed-rail-no-turning-back-on-american-jobs-economic-opportunities-mobility-gains.html. See also "High-Speed Rail Passenger Service Supported by Youngstown/Warren Regional Chamber," MfrTech.com, May 25, 2010, http://www.mfrtech.com/articles/3247.html (accessed October 8, 2010).

3. On metropolitan regionalism see, for example, Clyde Mitchell Weaver, David Miller, and Ronald Deal Jr., "Multilevel Governance and Metropolitan Regionalism in the United States," *Urban Studies* 37 (2000): 851–876; Peter Calthorpe and William Fulton, *The Regional City: Planning for the End of Sprawl* (Washington, DC: Island Press: 2001), 15–30. More recently, the scale of the economic urban region has come more closely to resemble Jean Gottman's idea of the megalopolis in *Megalopolis: The Urbanized Northeastern Seaboard of the United States* (New York: Twentieth Century Fund, 1961). See Robert E. Lang and Arthur C. Nelson, "The Rise of the Megapolitans," *Planning* 73, no.1 (2007): 7–12; Catherine L. Ross, "Megaregions: Literature Review of the Implications for U.S. Infrastructure Investment and Transportation Planning," report submitted by the Center for Quality Growth and Regional Development at the Georgia Institute of Technology to the U.S. Department of Transportation Federal Highway Administration (September 2008); and Ross (ed.), *Megaregions: Planning for Global Competitiveness* (Washington, DC: Island Press, 2009). For an enthusiastic discussion of megaregions (whose outlines differ slightly from other models), see Richard Florida, *The Great Reset: How New Ways of Living and Working Drive Post-Crash Prosperity* (New York: HarperCollins, 2010), 141–155. For a thoughtful and detailed three-part discussion of the idea of the megaregion, particularly as it applies to the Midwest, see Aaron Renn's The Urbanophile blog, July 11, 2008, December 6, 2009, and December 13, 2009. Renn reminds readers of this quote by Lewis Mumford, who, not surprisingly, found the notion alarming: "Instead of creating the Regional City, the forces that automatically pumped highways and motor cars and real estate development into the open country have produced the formless exudation. Those who are using verbal magic to turn this conglomeration into an organic entity are only fooling themselves. To call the resulting mass 'Megalopolis,' or to suggest that changes in spatial scale, with swift transportation, in itself is sufficient to produce a new and better urban form, is to overlook the complex nature of the city. The actual coalescence of urban tissue now taken by many sociologists to be a final stage in city development, is not in fact a new sort of city, but an anti-city. As in the concept of anti-matter, the anti-city annihilates the city whenever

it collides with it." See Mumford, *The City in History* (New York: Harcourt, 1961), 505.

4. Regional Plan Association, *America 2050: An Infrastructure Vision for the 21st Century* (2008).

5. Transportation for America, "Platform for the National Transportation Program Authorization" (March 2009), 5, 17.

6. See, for example, Lang and Nelson, "Rise of the Megapolitans."

7. Robert Yaro, Robert Wonderling, and other members of Business Alliance for Northeast Mobility to U.S. Department of Transportation Secretary Ray LaHood, June 22, 2010; Angela Cotey, "Ohio Rail Group Seeks to Advance Smaller Passenger-Rail Projects within 3C Corridor," Hsrupdates.com, April 25, 2011, http://www.hsrupdates.com/news/details/Ohio-rail-group-seeks-to-advance-smaller-passengerrail-projects-within-3C-corridor-871 (accessed June 18, 2011).

8. Raymond A. Mohl, "The Interstates and the Cities: Highways, Housing, and the Freeway Revolt" (Washington, DC: Poverty and Race Research Action Council, 2002).

9. Ibid., 28.

10. Wilbur Smith and Associates, *Future Highways and Urban Growth* (Automobile Manufacturers Association, February 1961), 67, 218.

11. Tom Lewis, *Divided Highways*, (New York: Penguin, 1997), 153. Onondaga Citizens League, "Rethinking I-81" (2009).

12. For a more sanguine study of urban freeway impact mitigation, notably the cost and effects of depressing these highways below grade, see Thomas B. Gray, "The Aesthetic Condition of the Urban Freeway," unpublished report for the Department of Community and Regional Planning, University of Texas at Austin, December 1999, especially chapter 3, http://www.mindspring.com/~tbgray/prindex.htm (accessed December 30, 2010).

13. Telephone interview with John O. Norquist, January 29, 2010.

14. Onondaga Citizens League, "Rethinking I-81"; telephone interview with Norquist, January 29, 2010; interview with Robert Simpson, August 24, 2009. See also John O. Norquist, *The Wealth of Cities: Revitalizing the Centers of American Life* (New York: Basic Books, 1999) and Congress for the New Urbanism, "Freeways without Futures," 2008, http://www.cnu.org/highways/freeways withoutfutures (accessed December 2008).

15. William Neuman, "Creativity Helps Rochester's Transit System Turn a Profit," *New York Times*, September 15, 2008.

16. Rochester-Genesee Regional Transportation Authority, news release, September 8, 2009.

17. See Wight and Company, Wolff Clements and Associates, and Gary W. Anderson and Associates, "West State Street Corridor Study," 2002. My characterization of the project, which has been altered over the past eight years, also reflects comments made to me by city officials during a meeting on November 10, 2009.

18. See Northern Illinois Commuter Transportation Initiative, "Alternative Analysis and Environmental Assessment Study," 2006, http://www.nicti.net/PO-Purpose%20&%20Need.htm (accessed February 2, 2010).

Chapter 3

1. The conference papers have been published in book form. See Mohsen Mostafavi and Gareth Doherty (eds), *Ecological Urbanism* (Baden, Germany: Lars Müller Publishers, 2010).

2. Patrick Geddes, "Civics as Applied Sociology," in *Sociological Papers*, V. V. Branford, ed. (London: MacMillan, 1905), 105.

3. For a reproduction of Geddes's 1905 Valley Section transect diagram, along with a more detailed, illustrated history of transect ideas from a new urbanist perspective, go to http://www.transect.org/natural_img.html. See also Ian McHarg, *Design with Nature* (Washington, DC: American Museum of Natural History, 1969). Inspired by the ecology movement, McHarg revived the concept of the transect, but, like Geddes, neglected to integrate it with urban form.

4. Andrés Duany, "Introduction to the Special Issue Dedicated to the Transect," *Journal of Urban Design* 7 (2002): 251–260; Duany and Emily Talen, "Making the Good Easy: The Smart Code Alternative," *Fordham Urban Law Journal* 29 (2002): 1445–1468. See also the work of the Form-Based Codes Institute, at www.formbasedcodes.org, and the Rocky Mountain Land Use Institute's "Sustainable Community Development Code Framework," forthcoming at the time this book went to press, which provides for more agriculture and renewable energy generation in urban settings than most form-based codes, at http://law.du.edu/index.php/rmlui/rmlui-practice/code-framework (accessed March 19, 2011). New urbanist aesthetics is indebted not only to the City Beautiful but also to the more recent work of Léon Krier, *The Architecture of Community*, ed. Dhiru A. Thadani and Peter J. Hetzel (Washington, DC: Island Press, 2009), and Christopher Alexander, with Sara Ishikawa and Murray Silverstein, *A Pattern Language: Towns, Buildings, and Construction* (New York: Oxford University Press, 1977); "A City Is Not a Tree," *Architectural Forum* 122 (April 1965): 58–62, (May 1965): 58–61. See also Emily Talen and Cliff Ellis, "Cities as Art: Exploring the Possibility of an Aesthetic Dimension in Planning," *Planning Theory and Practice* 5, no. 1 (March 2004): 11–32.

5. A month after the conference, the Harvard School of Design appointed Charles Waldheim, leading light of the emergent landscape urbanist movement, chair of the landscape architecture department and has since made ten landscape urbanist faculty appointments, thus demonstrating its embrace of the movement. As this book went to press, Waldheim—who has been quoted as saying that his program seeks "specifically" and "explicitly" to upend new urbanist influence—and Duany, along with their followers, were engaged in a pitched public battle, previewed at this conference, over urbanism's relationship to ecological systems. See Leon Neyfakh, "Green Building," *Boston Globe* (January 30, 2011) http://www.boston.com/bostonglobe/ideas/articles/2011/01/30/green_building/ (accessed March 15, 2011).

6. Duany Plater-Zyberk & Company, LLC, "Agricultural Urbanism," draft report, November 18, 2009. On further customizing form-based codes for weak-market cities, see Joseph Schilling and Jonathan Logan, "Greening the Rust Belt: A Green Infrastructure Model for Right Sizing America's Shrinking Cities," *Journal of the American Planning Association* 74, no. 4 (Autumn 2008): 462.

7. Charles C. Bohl with Elizabeth Plater-Zyberk, "Building Community across the Rural-to-Urban Transect," *Places* 18, no. 1 (2006): 15.

8. Telephone interview with Kevin Klinkenberg, February 5, 2010. See also Nate Berg, "Brave New Codes," *Architect* (July 2010), http://www.architectmagazine.com/codes-and-standards/brave-new-codes.aspx (accessed March 18, 2011).

9. Michael Pollan, *The Omnivore's Dilemma* (New York: Penguin, 2006), esp. 15–64. According to the National Academy of Sciences, corn ethanol generates only a 25 percent net gain in energy because it requires so many inputs. See Jason Hill et al., "Environmental, Economic, and Energetic Costs and Benefits of Biodiesel and Ethanol Biofuels," *Proceedings of the National Academy of Sciences of the United States of America* 103, no. 30 (2006): 11206–11210.

10. Wendell Berry, *The Unsettling of America: Culture and Agriculture* (New York: Avon, 1997). For a polemical tract opposing these views by one of Berry's erstwhile critics, see Blake Hurst, "The Omnivore's Delusion against the Agri-Intellectuals," *American* blog, July 30, 2009.

11. See J. S. Nettleton, "Regional Farmers' Market Development as an Employment and Economic Development Strategy," in *Proceedings of the Environmental Enhancement through Agriculture Conference,* ed. W. Lockeretz (Medford, MA: Center for Agriculture, Food and Environment, Tufts University, 1995), 235–243; Scott M. Springer, "FoodNYC: A Blueprint for a Sustainable Food System" (New York: New York University and Just Food, February 2009).

12. Interview with Henry Brockman, November 8, 2010.

13. Henry Brockman, *Organic Matters* (Congerville, IL: TerraBooks, 2001), 4.

14. See also Bill McKibben, *Deep Economy: The Wealth of Communities and the Durable Future* (New York: Holt, 2007), and *Eaarth: Making a Life on a Tough New Planet* (New York: Times Books, 2010).

15. Telephone interview with Terra Brockman, October 28, 2008. The argument that globalization undermines efforts to refigure the agricultural economy along more sustainable, local lines is made forcefully, if sympathetically, by Richard Longworth, *Caught in the Middle* (New York: Bloomsbury, 2008), 62–81.

16. National Oceanic and Atmospheric Administration, "Smaller Than Expected, But Severe, Dead Zone in Gulf of Mexico," July 27, 2009. For good summaries of the growing literature (including several recent studies by the United Nations) showing that small local freeholders practicing sustainable agriculture can meet the demand for food in impoverished countries plagued by hunger, see UN Conference on Trade and Development and UN Environment Programme Capacity Building Task Force, *Organic Agriculture and Food Security in Africa* (New York: United Nations, 2008), http://www.unctad.org/en/docs/ditcted200715_en.pdf (accessed October 9, 2010); UK HungerFree Campaign,

"Brief on Sustainable Agriculture" (October 2009), http://www.actionaid.org.uk/doc_lib/sustainable_agriculture_aa.pdf (accessed October 10, 2010); Eric Hewlett and Peter Melchett, "Can Organic Agriculture Feed the World? A Review of the Research," paper delivered at the IFOAM Organic World Congress, Modena, Italy, June 16–20, 2008, http://www.pigbusiness.co.uk/pdfs/Soil-Association-Can-Organic-feed-the-World.pdf (accessed October 10, 2010). For a hearty defense of the traditional "feed the world" argument, see Robert Paarlberg, "Attention, Whole Food Shoppers," *Foreign Policy* (May–June 2010), http://www.foreignpolicy.com/articles/2010/04/26/attention_whole_foods_shoppers (accessed October 10, 2010).

17. See Mike Hamm, "Michigan Economic Development: Opportunities through Local Sustainable Agriculture," presentation to the Greentown Conference, Grand Rapids, MI, July 23, 2009, http://wn.com/Green_Town_Grand_Rapids (accessed February 8, 2010). Much of this talk is based on a coauthored article: David S. Conner et al., "The Food System as an Economic Driver: Strategies and Applications for Michigan," *Journal of Hunger and Environmental Nutrition* 3 (2008): 371–383.

18. See Ken Meter and Jon Rosales, "Finding Food in Farm Country: The Economics of Food and Farming in Eastern Minnesota" (University of Minnesota, 2001), updated through 2003 at http://www.crcworks.org/fffc.pdf (accessed February 15, 2010).

19. Interview with Mike Hamm, September 15, 2009. Alisa Smith and J. B. MacKinnon, *Plenty: One Man, One Woman, and a Raucous Year of Eating Locally* (New York: Harmony/Random House, 2007).

20. See Michigan Food Policy Council, "Report on Recommendations," October 12, 2006, http://www.michigan.gov/documents/mda/MFPC_Report_2006_174216_7.pdf (accessed February 15, 2010). For more on food policy councils, see Althea Harper et al., *Food Policy Councils: Lessons Learned* (Oakland, CA: Institute for Food and Development Policy, 2009), http://www.foodfirst.org/files/pdf/Food%20Policy%20Councils%20Report%20small.pdf (accessed February 15, 2010), and Kate Clancy et al., "Food Policy Councils: Past, Present, and Future," in *Remaking the North American Food System*, ed. C. Clare Hinrichs and Thomas A. Lyson (Lincoln, Nebraska: University of Nebraska Press, 2007), 121–140.

21. Interview with Patty Cantrell, September 17, 2009. For a good overview of the challenges school lunch programs face in sourcing local food, see Val George, Colleen Matts, and Susan Schmidt, *Institutional Food Purchasing: Michigan Good Food Work Group Report No. 3 of 5* (East Lansing, MI: C. S. Mott Group for Sustainable Food Systems at Michigan State University, 2010), http://www.michiganfood.org/assets/goodfood/docs/Inst%20Food%20Purchasing%20Report.pdf (accessed December 30, 2010).

22. Jason DeParle and Robert Gebeloff, "Food Stamp Use Soars, and Stigma Fades," *New York Times*, November 29, 2009, A-1.

23. Suzanne Briggs et al., "Real Food, Real Choices: Connecting SNAP Recipients with Farmers Markets (San Francisco: Tides Foundation, June 2010), esp. 68,

http://www.foodsecurity.org/pub/RealFoodRealChoice_SNAP_FarmersMarkets.pdf (accessed January 1, 2011). "Michigan Ranks Fourth in Nation for Farmers Markets," Govmonitor.com, August 5, 2010, http://www.thegovmonitor.com/world_news/united_states/michigan-ranks-fourth-in-nation-for-farmers-markets-36600.html (accessed January 1, 2011).

24. Nancy Rosin, *The Hands That Feed Us* (Rochester: City of Rochester, New York, 2005), 64.

25. This study asks a similar question of six large metro areas: Francesca Pozzi and Christopher Small, "Vegetation and Population Density in Urban Suburban Areas in the U.S.A.," paper presented at the Third International Symposium of Remote Sensing of Urban Areas, Istanbul, Turkey, June 2002, http://sedac.ciesin.columbia.edu/urban_rs/PozziSmall2002.pdf (accessed February 16, 2010). See also Reid Ewing, Rolf Pendell, and Don Chen, *Measuring Sprawl and Its Impact* (Washington, DC: Smart Growth America, 2002), 30, which observes, "the largest metro areas, perceived as the most sprawling by the public, actually appear less sprawling than smaller metros when sprawl is measured strictly in terms of the four factors, with no consideration given to size." At least part of the answer may lie buried in USDA Economic Research Service figures. See its "Measuring Rurality: Urban Influence Codes," 2003, http://www.ers.usda.gov/Briefing/rurality/UrbanInf/ (accessed February 20, 2010). My own inexpert crunching of 2000 Census data bears out the notion that smaller cities are less densely settled, closer in. In suburban Peekskill, New York, 41 miles north of New York City, the population density is 5,189; in New Rochelle, 14 miles away, density is 6,973. Here are figures for suburban Rochester: Churchville (14 miles, 1,662), Webster (10 miles, 1,114), Fairport (10 miles, 3,645), Irondequoit (4 miles, 3,447), and Chili (adjacent, 695). Moving to the Midwest, here are figures for suburban Chicago: Downers Grove (21, miles, 3,420) and Elgin (36 miles, 3,779). And here are figures for suburban Peoria: East Peoria (adjacent, 1,203), Morton (12 miles, 1,247), Washington (10 miles, 1,450), and Germantown (7 miles, 578). See http://www.census.gov/population/www/censusdata/density.html (accessed February 21, 2010). Of course, none of this takes into account how much of that land is covered with commercial development and roads, much less residential housing. What geographers and economists calls "land cover," gathered through GIS technology by the USDA and National Oceanic and Atmospheric Administration, would provide more meticulous data, but no one to date has analyzed these data with small-metro-area suburban density in mind. Agricultural economist Christian Peters has used this data to map potential foodsheds in upstate New York, showing that there is enough available fertile land to feed the entire state outside metropolitan New York City. He is now working, with colleagues on a much larger land-cover mapping project extended to the rest of the country. See Christian J. Peters et al., "Mapping Potential Foodsheds in New York State: A Spatial Model for Evaluating the Capacity to Localize Food Production," *Renewable Agriculture and Food Systems* 21, no. 1 (2009): 72–84.

26. Ralph E. Heimlich and William D. Anderson, *Development at the Urban Fringe and Beyond: Impacts on Agriculture and Rural Land*—AER-803, USDA

ERS (2001): vi, 41–42; "Land Use, Value, and Management: Urbanization and Agricultural Land," USDA ERS briefing, June 28, 2005.

27. On the new ruralism, see Sibella Kraus, "A Call for New Ruralism," *Frameworks* (spring 2006): 26–29; David Moffat, "New Ruralism: Agriculture at the Metropolitan Edge," *Places* 18, no. 2 (Fall 2006): 72–75; and Emily M. Stratton, *New Ruralism*, report for the University of Georgia Land Use Clinic (Fall 2009). See also Duany Plater-Zyberk & Company, LLC, *Agricultural Urbanism*, draft report, November 18, 2009. On conservation subdivisions, see Randall Arendt, *Conservation Design for Subdivisions: Guide to Creating Open Space Networks* (Washington, DC: Island Press, 1996) and *Growing Greener: Putting Conservation into Local Plans and Ordinances* (Washington, DC: Island Press, 1999). See also Kimberly Bosworth, "Conservation Subdivision Design: Perceptions and Reality," master's thesis, University of Michigan, 2007, and Rayman Mohamed, "The Economics of Conservation Subdivisions," *Urban Affairs Review* 41 (2006): 376–399. Duany's comments were made in "Introduction to the Special Issue Dedicated to the Transect," 3. These developments are descended, to varying degrees, from Frank Lloyd Wright's early 1930s concept of the suburban Broadacre community, which allocated acre lots to each house, intended for individual agricultural use, and tied together by automobile-accommodating roads.

28. Interview with Virginia Ranney, November 11, 2009.

29. Urban Land Institute, "Case Study: Prairie Crossing, Grayslake, Illinois," in *Developing Sustainable Planned Communities* (Washington, DC: ULI, 2007), 209. In 1999, George Ranney told the *New York Times* that 8 percent of Prairie Crossing residents were African American, in a town where the average was 1 percent. "Developing a Suburb, with Principles, *New York Times*, July 11, 1999.

30. On the Madison project, see http://www.conservationfund.org/node/693. On Fisher Creek, see http://www.fishercreekneighborhood.com/our-vision/. See also http://savannahnow.com/dahleen-glanton/2009-03-07/serenbe-community-palmetto-ga-plunges-green-living; http://www.southvillage.com/news004.php; http://friendofthefarmer.com/2009/08/organic-farm-replaces-golf/ (all accessed February 18, 2010). Ed McMahon, *Conservation Communities: Creating Value with Nature, Open Space, and Agriculture* (Washington, DC: Urban Land Institute, 2010).

31. On my own minimal involvement in this longer legal battle, see Brian Rooney, "Army Engineers Sued in Attempt to Block Mall," *Rochester Democrat and Chronicle*, April 24, 1980, B-3, and Dolores Orman, "Suit to Stop Marketplace Going to Trial," *Rochester Times-Union*, May 7, 1980. On the legal dispute in general, see Marketplace Mall clippings file, Local History Department, Central Library of Rochester and Monroe County. For a good general overview of the battle, see Mark Wert, "Mall to Be Built after a Long Wait, Many Disputes," *Rochester Democrat and Chronicle*, December 7, 1970, B-8. By executive order, President Jimmy Carter put into effect requirements for an urban impact analysis for major federal projects in 1978, but it was not retained by the Reagan administration. See Thomas Kingsley and Karina Fortuny, "Urban Policy in the Carter Administration," May 2010, 4, http://www.urban.org/uploadedpdf/412091-carter-urban-policy.pdf (accessed January 1, 2011).

32. Karen McCally, "The Life and Times of Midtown Plaza," *Rochester History* 69:1 (2007), http://www.rochesterhistory.org/documents/roch_history_magazine.pdf (accessed March 2, 2010).

33. See Dolores Hayden, *Building Suburbia: Greenfields and Urban Growth, 1820–2000* (New York: Vintage, 2003), 162–180, 231. On land lease and deed arrangements, see Julia Christensen, *Big Box Reuse* (Cambridge, MA: MIT Press, 2008), 8–9, 38–40. See also Ellen Burnham-Jones and June Williamson, *Retrofitting Suburbia: Urban Design Solutions for Redesigning Suburbs* (New York: Wiley, 2008).

34. Christensen, *Big Box Reuse*, 5. On retail development on wetlands and creek beds, see Burnham-Jones and Williamson, *Retrofitting Suburbia*, 114.

35. Nate Berg, "Redressing Strip Malls," Planetizen blog, November 14, 2008, http://www.planetizen.com/node/36042 (accessed February 25, 2010).

36. Christensen, *Big Box Reuse*, 5; Sheryl Rich-Kern, "Secondhand Shops Supersize to Maximize Potential," National Public Radio Morning Edition, March 3, 2010, http://www.npr.org/templates/story/story.php?storyId=124202298 (accessed March 5, 2010).

37. Burnham-Jones and Williamson, *Retrofitting Suburbia*, 158–171.

38. Stephanie Kist, "Developer Plans Redevelopment of State Road Shopping Center," Akron.com, May 21, 2009; Emily Chesnik, "Falls Approves Design Changes to State Road, Portage Trail," Akron.com, March 25, 2010; Cuyahoga Falls City Council minutes, January 4, 2010, http://cfo.cityofcf.com/content/council/2010/1-4-10%20Finance.pdf (accessed October 12, 2010).

39. Although green efficiency can lead to substantial savings in electricity, water, and other operating costs, they're passed on to tenants, who are responsible for these expenses through standard "triple net" leases, which require tenants to pay a proportionate share of insurance, taxes, and maintenance costs, as well as rent.

40. According to the ICSC's sustainability portal Web page, February 22, 2010, "Kohl's Department Stores, Menomonee Falls, WI, has passed PepsiCo to become the nation's No. 2 user of green power, and is closing in on U.S. leader Intel Corp. Whole Foods Markets came in at No. 4. Kohl's has been a member of EPA's Green Power Partnership since 2006 and was named a 2009 Green Power Partner of the Year. Kohl's was also honored in 2007 and 2008 with Green Power Leadership Awards for green power purchase and on-site generation. In addition, the company's green initiatives were recognized in 2009 with an EPA SmartWay Excellence award and a ranking of No. 1 in retail on *Newsweek*'s Green Rankings." See also Doris Hajewski and Don Walker, "Kohl's to Close Menomonee Falls Distribution Center," *Milwaukee Journal Sentinel*, October 13, 2009. On green building practices in the retail industry, see Jerry Yudelson, "LEEDing Retail to Greener Pastures," *Research Review* 16:2 (2009), 54–60, and Robert J. Sykes, "2009 ICSC RetailGreen Conference: There's Green in Going Green," Cox Castle Client Alert, December 2009.

41. Joel Garreau, "Today's Temples of Consumption Don't Have to Be Tomorrow's Ruins. What's in Store?" *Washington Post*, November 16, 2008, M-1; "They Say It Doesn't Grow on Trees. Wanna Bet?" *Washington Post,* November 16,

2008, B-4; Forrest Fulton, "Big Box Agriculture: A Productive Suburb," third-place winner in the ReBurbia Design Competition sponsored by *Dwell* magazine and Inhabitat Web site, August 19, 2009, http://www.re-burbia.com/2009/08/01/a-new-business-model-a-productive-suburb/ (accessed March 1, 2010).

42. Lisa Selin Davis, "The Secret Mall Gardens of Cleveland," *Grist*, March 17, 2010, http://www.grist.org/article/the-mall-gets-fresh (accessed October 15, 2010). Ariel Schwartz, "At Cleveland Mall Green Market, Sustainability Is the Hot New Topic," *Fast Company*, March 9, 2010, http://www.fastcompany.com/1576976/cleveland-galleria-mall-greenhouse-gardens-under-glass (accessed October 15, 2010).

43. Bonnie Azab Powell, "The New Agtivist: Gene Fredericks Is Thinking Inside the City's Big Box," *Grist*, September 1, 2010, http://www.grist.org/article/food-the-new-agtivist-gene-fredericks-is-thinking-inside-the-citys-bi (accessed October 16, 2010).

44. Marc Gunther, "'An Emotional, Economic, Social Reset,'" marcgunther.com, November 6, 2008, http://www.marcgunther.com/2008/11/06/an-emotional-social-economic-reset/ (accessed March 5, 2010).

Chapter 4

1. Novella Carpenter, *Farm City: The Education of an Urban Farmer* (New York: Penguin, 2009).

2. David Owen, *Green Metropolis: Why Living Smaller, Living Closer, and Driving Less Are the Keys to Sustainability* (New York: Riverhead, 2009), 2–3. See also my review: "The City's Limits," *Wilson Quarterly* 34, no. 1 (Winter 2010): 108–109.

3. Some 20 million Americans participated in the effort, which at, its peak, produced as much as 40 percent of domestic vegetable produce. See Emanuel B. Halper, *Shopping Center and Store Leases*, vol. 2 (Brooklyn, NY: Law Journal Seminars Press, 1979), 93–99; Kenneth Helphand, *Defiant Gardens: Making Gardens in Wartime* (San Antonio, TX: Trinity University Press, 2006); Laura Lawson, *City Bountiful: A Century of Community Gardening in America* (Berkeley: University of California Press, 2005); and David J. Hess, *Localist Movements in a Global Economy* (Cambridge, MA: MIT Press, 2009), 138–156. Depression-era work relief gardens are another antecedent, though they have received less historical study. For a brief overview, with references, see L. E. Heimer, "Depression Relief Gardens," sidewalksprouts.wordpress.com, n.d., http://sidewalksprouts.wordpress.com/history/relief-garden/ (accessed January 2, 2011).

4. In 2010, the German Marshall Fund launched a three-year program facilitating transatlantic conversation among specialists and policymakers about "cities in transition." See http://www.gmfus.org/cs/citiesintransition (accessed March 28, 2011). See also Terry Schwarz, "The Cleveland Land Lab: Experiments for a City in Transition," in *Cities Growing Smaller*, ed. Steve Rugare and Terry Schwarz (Cleveland: Cleveland Urban Design Collaborative, 2008), 72–83, which provides a clear and accessible exposition of what this new urban form might look like.

On urban ecological restoration, see Joan Iverson Nassauer (ed.), *Placing Nature: Culture and Landscape Ecology* (Washington, DC: Island Press, 1997); Charles Lord, Eric Strauss, and Aaron Toffler, "Natural Cities: Urban Ecology and the Restoration of Urban Ecosystems," *Virginia Environmental Law Journal* 21 (2003): 454–551. For a comprehensive policy approach to the problem of shrinking cities in Ohio, one that includes recommendations for urban agriculture and green infrastructure, see Alan Mallach and Lavea Brachman, *Ohio Cities at a Turning Point: Finding a Way Forward* (Washington, DC: Brookings Institution, 2010).

5. See, for example, a blog debate between Roberta Brandes Gratz and Allan Mallach, "Demolition a Wrong Answer for Imperiled Neighborhoods," Citiwire.net, June 18, 2009, http://citiwire.net/post/1007/ (accessed June 24, 2009); Kaid Benfield, "They Are Stardust. They Are Golden. But Are They Right about 'Shrinking Cities'?" Switchboard.nrdc.org, July 2, 2009, http://switchboard.nrdc.org/blogs/kbenfield/they_are_stardust_they_are_gol.html (accessed October 5, 2009).

6. In Flint, a group of business leaders leveraged $200 million for downtown development, beginning in 2004. See Charles Stewart Mott Foundation, *"Flint: A Snapshot in Time": Charles Stewart Mott Foundation 2008 Annual Report* (Flint, MI: Charles Stewart Mott Foundation, 2008), p. 14.

7. Interview with Dan Kildee, September 14, 2009. For an excellent and comprehensive overview of how the Genesee County Land Bank came into being and how it works, see Nigel G. Griswold and Patricia E. Norris, "Economic Impacts of Residential Property Abandonment and the Genesee County Land Bank in Flint, Michigan," MSU Land Policy Institute Report 2007–05 (April 2007), esp. 19–26. See also Chris McCarus, "Banking on Flint: County Treasurer Dan Kildee Collects National Attention for Land Bank," *Dome*, July 16, 2008; Steve Dubb, interview with Dan Kildee, Community-Wealth.org, June 2009, http://www.community-wealth.org/_pdfs/news/recent-articles/07-09/interview-kildee.pdf (accessed September 2, 2009).

8. Steve Inskeep, "Flint, Michigan: Growing Stronger by Growing Smaller?" NPR, *Morning Edition*, July 13, 2009, http://www.nprinternedition.org/templates/transcript/transcript.php?storyId=106492824.

9. Interview with Harry Ryan, September 14, 2009. At the time of my visit, 28 percent of Flint residents were unemployed, and 35 percent lived below the poverty line.

10. One of the chief reasons—in addition to the intransigence of the mayor—that Buffalo can't seem to avert its slide in population and jobs is that no family foundation is present to target nongovernment funding for community development and long-range planning. My thanks to Joe Schilling, of the Virginia Tech Metropolitan Institute and the National Vacant Properties campaign, for pointing this out in a comment on an earlier draft of this chapter.

11. Telephone interview with Roxanne Adair, March 8, 2010.

12. Interview with Roxanne Adair, September 14, 2009.

13. Flint Futures Group, *Reimagining Chevy in the Hole* (Ann Arbor: School of Natural Resources and Environment, University of Michigan, 2007).

14. Dave Hoekstra, "Rebuilding Flint: A Desolate Town Driven to Be Green," *Shore*, March 19, 2010.

15. Mark Dowie, "Food among the Ruins," *Guernica*, August 2009 http://www.guernicamag.com/features/1182/food_among_the_ruins/ (accessed March 9, 2010).

16. Mallach, Alan, et al. *Leaner, Greener Detroit* (Washington, DC: American Institute of Architects, Sustainable Design Assessment Team, 2008), 7. This initiative was highly unusual. SDAT studies are usually commissioned directly by municipalities and include careful efforts to hear from a variety of urban constituencies. This study of Detroit, conducted on a more private basis, has been criticized for not covering all the political bases. See also Nancy Kaffer, "Some Say City Could Do More with Less," *Crain's Detroit Business*, August 23, 2009.

17. For a more complete breakdown of the obstacles facing Bing's resizing plans, including legal challenges, see Christine MacDonald, "Detroit Mayor Bing Emphasizes Need to Shrink City," *Detroit News*, February 25, 2010, and Alex P. Kellogg, "Detroit's Smaller Reality: Mayor Plans to Use Census Tally Showing Decline as Benchmark in Overhaul," *Wall Street Journal*, February 27, 2010.

18. These tensions are spelled out in more detail in Minehaha Forman, "Race Dynamic Seen as Obstacle in Detroit Urban Farming," *Michigan Messenger*, October 30, 2009.

19. David Whitford, "Can Farming Save Detroit?" *Fortune*, 161, no. 1 (January 18, 2010): 78–84. See also Jason A. King, "The Detroit Dilemma-Ruminations," Landscape + Urbanism blog, December 26, 2008. King participated in the American Institute of Architects' Sustainable Design Assessment Team study.

20. Jonathan Oosting, "Will Urban Farming Save Detroit? All Eyes on John Hantz," MLive.com, January 4, 2010.

21. Daniel Okrent, "The Tragedy of Detroit: How a Great City Fell—and How It Can Rise Again: A Special Report," *Time*, October 5, 2009, 26–35. See also the PBS documentary *Blueprint America: Beyond the Motor City* (2010), and Bruce Katz and Jennifer Bradly, "The Detroit Project: A Plan for Solving America's Greatest Urban Disaster," *New Republic*, December 2, 2009, 29–31.

22. Greg Lindsay, "Demolishing Density in Detroit: Can Farming Save the Motor City?" *Fast Company* blog, March 4, 2010. He also argues that Detroit is already dense compared with other American cities, but ignores the fact that the population is more evenly distributed elsewhere.

23. See Herbert Gans, *The Urban Villagers: Group and Class in the Life of Italian Americans* (New York: Macmillan, 1962).

24. Douglas Belkin, "Undecideds Think It Over," *Wall Street Journal*, September 29, 2008.

25. Seth Teter, "A Shrimp Tale: How Ohio Came to Produce Fresh Seafood," *Our Ohio* (January 2006); Fran Henry, "Ohio's Farmers Expand into Shrimp Aquaculture," *Cleveland Plain Dealer,* August, 31, 2005, http://www.usmsfp.org/news/headlinenews/08-31-2005-ohioshrimpfarms.htm (accessed March 4, 2010).

26. According to the U.S. Census, Youngstown's population dropped from 168,330 in 1950 to 66,982 in 2010, bearing out an earlier claim that it tops the list of U.S. cities with the biggest population losses. See Lauren Sherman, "America's Downsized Cities," *Forbes*, March 18, 2010.

27. Tony Favro, "American Cities Seek to Discover Their Right Size," Citymayors.com, April 5, 2010, http://www.citymayors.com/development/us-rightsizing-cities.html (accessed April 10, 2010).

28. Interview with Bill D'Avignon, October 19, 2009.

29. Interview with Ian Benisten, October 18, 2009. On some of the challenges the city has had in implementing Youngstown 2010, see Dan Kildee et al., "Regenerating Youngstown and Mahoning County through Vacant Property Reclamation: Reforming Systems and Right-sizing Markets," National Vacant Properties Campaign Policy Assessment Report (February 2009).

30. George Nelson, "Group Targets 3 'Healthy Neighborhoods,'" Business-journal.com, February 9, 2010. For another proposal to incorporate green infrastructure into Youngstown's comprehensive land use plan, see John Bralich, "Developing Methods to Establish an Urban Wetland Mitigation Bank on Youngstown's East Side," draft report prepared for the City of Youngstown (December 2009). On repurposing urban land in shrinking cities for ecological restoration and urban agriculture, see Rugare and Schwarz (eds.), *Cities Growing Smaller*, and Joe Schilling and Jonathan Logan, "Greening the Rust Belt: A Green Infrastructure Model for Right Sizing America's Shrinking Cities," *Journal of the American Planning Association* 74 (2008): 451–466.

31. Telephone interview with Elsa Higby, March 20, 2010. See also Denise Dick, "Growing for Change: Garden Brings Out Goodness in Community," Vindy.com, October 2, 2009; Elimy Caldwell, "Ohio Church Serves God, Community with Small-Scale Urban Gardening," Farmanddairy.com, July 9, 2009; Sudhir Kade, "The Great Will Allen on Urban Agriculture at the Grey to Green Festival in Youngstown," Realneo.us, October 5, 2009.

32. Telephone interviews with Maurice Small, March 25, 2010 and March 2, 2011.

33. Interview with Phil Kidd, October 21, 2009. These figures are included in a January 20, 2009, memo by Ian Beniston posted on MVOC's Web site: http://www.mvorganizing.org/downloads/MVOC_Strategic_Demolition_Policy_Brief.pdf (accessed March 10, 2010).

34. See Jim Russell, "Destinations of Out-Migrants from 5-County Youngstown Region by CBSA, 1990–2007," Greater Youngstown 2.0 Project, http://globalburgh.com/gytwodotzero/Youngstown_map_Nation_1.jpg (accessed March 11, 2010).

35. Gary Painter and Zhou Yu, "Immigrant Influx into Mid-size Metro Areas to Influence Housing Markets," USC Lusk Center for Real Estate (Spring 2010).

36. Alan Ehrenhalt, "Trading Places: The Demographic Inversion of the American City," *New Republic*, August 13, 2008; Audrey Singer, "The New Geography

of United States Immigration," Brookings Institution Immigration Series Report (July 2009).

37. Radhika Fox and Miriam Axel-Lute, *To Be Strong Again: Renewing the Promise in Smaller Industrial Cities* (New York: PolicyLink, 2008), 21.

38. Beth Siegel and Amy Waxman, *Third Tier Cities: Adjusting to the New Economy*, Reviews of Economic Development Literature and Practice, no. 6 (Washington, DC: U.S. Economic Development Corporation, 2001).

39. U.S. Census, 1990 and 2000.

40. Interview with Geraldo Ramos, July 2008. See also Patrick O'Connor, "Coral Reef Grows in the Urban Jungle," Masslive.com, September 10, 2009; Corby Kummer, "A Papaya Grows in Holyoke," *Atlantic Monthly* (April 2008); Robert Gottlieb and Anupama Joshi, *Food Justice* (Cambridge, MA: MIT Press, 2010), 123–126.

41. See Northeast Network of Immigrant Farming Projects' Web site: http://www.immigrantfarming.org/_webapp_2767313/Northeast_Network_of_Immigrant_Farming_Projects (accessed April 2, 2010); Alex Peshkov, "Holyoke Farm Program Aids Immigrants," *Springfield Republican*, February 23, 2008.

42. Telephone interview with Daniel Ross, March 26, 2010.

43. On Nuestra Raíces's economic impact on Holyoke as of 2007, see Kay Oehler, Stephen C. Sheppard, and Blair Benjamin, "The Economic Impact if Nuestras Raíces on the City of Holyoke: Current and Future Projections" (Williamstown, MA: Center for Creative Community Development, February 2007).

44. Bill Baue, "A Hybrid Model: Co-op and Nonprofits Launch Energy-Efficiency Company," *Sea Change* radio, January 20, 2010; Office of U.S. Congressman John Olver, "Olver Announces $540,000 Funding for Holyoke's Nuestras Raices," October 14, 2009.

Chapter 5

1. Joan Fitzgerald, *Emerald Cities: Urban Sustainability and Economic Development* (New York: Oxford University Press, 2010).

2. Richard Longworth, *Caught in the Middle: America's Heartland in the Age of Globalism* (New York: Bloomsbury, 2008), 44–45.

3. See Peter Drucker, *The Effective Executive* (New York: Harper & Row, 1966), which introduced the term; Daniel Bell, *The Coming of Post-Industrial Society: A Venture in Social Forecasting* (New York: Harper Colophon, 1974); and Robert Reich, *The Work of Nations: Preparing Ourselves for 21st Century Capitalism* (New York: Knopf, 1992).

4. Kevin Phillips, *Bad Money: Reckless Finance, Failed Politics, and the Global Crisis of American Capitalism* (New York: Viking, 2008); William K. Tabb, "The Centrality of Finance," *Journal of World-Systems Research* 13, no. 1 (2007): 2.

5. See Robert S. and Helen Merrell Lind, *Middletown: A Study in Modern American Culture* (New York: Harcourt, Brace, 1929), 10–17.

6. For local media coverage, see "Brevini Invests in Delaware County, Brings Hundreds of Jobs," Insideindianabusiness.com, November 18, 2008, http://www.insideindianabusiness.com/advanced-manufacturing.asp?detail=true&id=96 (accessed April 25, 2010); "Delaware County Rail Spur Expected to Attract Investment," Insideindianabusiness.com, October 15, 2009, http://www.insideindianabusiness.com/advanced-manufacturing.asp?detail=true&id=186 (accessed April 20, 2010).

7. The $14 hourly wage estimate comes from historian James Connolly, who teaches at Ball State and produced a film documentary, *Changing Gears* (2010), on the closing of BorgWarner. E-mail to the author, June 15, 2010. See also "Neighbors Fret over Rail Plans for Turbine Plant," indy.com, March 5, 2009, http://www.indy.com/posts/neighbors-fret-over-rail-plans-for-turbine-plant (accessed April 29, 2010); "Wind Turbine Factory Coming to Muncie," *Indiana Railroads Bull Session: Today's Railroading: News*, October 8, 2008, http://indianarailroads.org/board/index.php?topic=2213.0 (accessed April 27, 2010); "Indianapolis Region Renewable Energy Components Manufacturing," Indy Partnership presentation, April 15, 2010, http://www.indypartnership.com/media/docs/clean%20tech%20docs/Clean-Tech_Presentation.pdf (accessed April 28, 2010).

8. "Wind Farm Developers Eye Muncie Area," indy.com, December 18, 2009, http://www.indy.com/posts/wind-farm-developers-eye-muncie-area (accessed April 27, 2010).

9. For a more detailed description of these strategies, see Fitzgerald, *Emerald Cities*, 14–16. On Wind Energy Manufacturer's Association, Inc., see http://www.inwema.org (accessed March 21, 2011) and http://www.muncie.com/Wind-Energy-Manufacturing-Association-WEMA.aspx (accessed March 19, 2011).

10. An antecedent to the principle of an RES is the national fuel standard written into the U.S. Energy Policy Act of 2005. It mandated the production of 7.5 billion gallons of renewable fuels by 2012 and was intended mainly as a boon to the corn ethanol industry. American Wind Energy Association, "Indiana Celebrates Wind Energy Week with Two Wind Farm Ground Breakings and #1 Wind Power Growth Rate Ranking," April 12, 2009, http://www.windfair.net/press/6051.html (accessed January 2, 2011); Lori Bird et al., *Evaluating Renewable Portfolio Standards and Carbon Cap Scenarios in the U.S. Electric Sector,* Technical Report NREL/TP-6A2–48258I (Golden, CO: U.S. Department of Energy, Office of Energy Efficiency and Renewable Energy, May 2010).

11. "Renato Brevini Wins the 'China Awards' 2008," *Brevini Power Transmission*, November 28, 2008, http://www.brevini.com/press00.asp?ID=393 (accessed April 9, 2010). It would be useful to undertake a comparative study of how smaller industrial cities, particularly in Europe, have transitioned to clean-tech industries that are now doing business with their U.S. counterparts. For a broad overview of what Detroit might learn from Turin, Italy (both large auto manufacturing cities), see Bruce Katz and Jennifer Bradley, "The Detroit Project," *New Republic*, December 9, 2009.

12. International Organization of Motor Vehicle Manufacturers, "2009 Production Statistics," www.oica.net, n.d., http://www.oica.net/category/production-statistics/ (accessed April 29, 2010).

13. Alisa Priddle, "Cracks in Auto Supply Chain Spreading," *Detroit News*, March 10, 2009.

14. The term *Auto Alley* itself wasn't coined until 2008. See Thomas Klier and James Rubenstein, *Who Really Made Your Car?* (Kalamazoo, MI: Upjohn Institute, April 2008). Also note that in the absence of media coverage of the recession's affect on the geography of the auto industry, columnist Paul Krugman felt compelled to draw attention to it in a 2009 *New York Times* blog post: Paul Krugman, "Black States," *New York Times*, August 3, 2009, http://krugman.blogs.nytimes.com/2009/08/03/black-states/ (accessed April 29, 2010).

15. Thomas Sugrue, *The Origins of the Urban Crisis: Race and Inequality in Postwar Detroit* (Princeton, NJ: Princeton University Press, 1996), 127–143.

16. Hudson Consulting, LLC, "The Economics and Politics of Smokestack Chasing," white paper (2003).

17. Sugrue, *Origins of the Urban Crisis*, 128, 141. On the de facto national industrial policy created by Pentagon contracting, see Robert Kuttner, *Everything for Sale: The Virtues and Limits of Markets* (New York: Knopf, 1997), chap. 6.

18. James M. Rubenstein and Thomas Klier, "Restructuring of the Auto Industry," paper delivered at the Industry Studies Association annual conference, Chicago, May 2009.

19. U.S. Department of Labor, Bureau of Labor Statistics, "Motor Vehicle and Parts Manufacturing," *Career Guide to Industries, 2010–2011 Edition*, modified date: December 3, 2010, http://www.bls.gov/oco/cg/cgs012.htm. On green cars, see David Landsman, "The Automotive Supply Chain: Glass Half-Full," Sourcing.community.mfg.com, January 16, 2010 (accessed May 2, 2010).

20. See "First Solar Overview,"http://www.firstsolar.com/Downloads/pdf/Fast Facts_PHX_NA.pdf (accessed March 9, 2011); James Kanter, "First Solar Claims $1-a-Watt 'Industry Milestone,'" *New York Times*, February 24, 2009. It's extremely difficult to calculate electricity use averages and prices (and impossible to determine generally how many homes can be powered by a megawatt—or, for that matter, a kilowatt—since prices are based on kilowatt hours and kilowatts reflect power intensity at any given moment. Generating plants also operate at various capacities, none at 100 percent, plus one must take into account variable energy use by region: the southern states, for example, use more energy due to heavy air-conditioning use. According to the Department of Energy, "Electricity Consumption by 107 million U.S. households in 2001 totaled 1,140 billion kWh." On average, that's 1,200 megawatts for 10,000 homes. This is the standard I am using here and throughout, distorted as it is. See U.S. Energy Information Administration, "Regional Energy Profiles: U.S. Household Electricity Report," July 14, 2005, www.eia.doe.gov (accessed April 22, 2010)). For a good overall description, see Bob Bellemare, "What Is a Megawatt?" *Utilipoint Issue Alert*, 2003.

21. Kevin Bullis, "U.S. Solar Market to Double in the Next Year," *Technology Review,* February 8, 2010. For a more detailed case study on which this discussion relies, see Fitzgerald, *Emerald Cities*, 56–59. On IMPACT, see Wendy Patton with analysis by Heidi Garrett-Peltier, "The Impact of IMPACT: Creating Jobs in Ohio," *Policy Matters* (February 2010).

22. Larry Ledebur and Jill Taylor, *Akron, Ohio: A Restoring Prosperity Case Study* (Washington, DC: Brookings Institution, 2007), 10–14.

23. Pete Engardio, "In Detroit, Is There Life after the Big 3?" *New York Times,* February 14, 2010; Dan Gearino, "Auto Industry's Pain Multiplied for Parts Suppliers," *Columbus Dispatch*, June 3, 2009.

24. On wind power alone, for example, see Gloria Ayee, Marcy Lowe, and Gary Gereffi, *Manufacturing Climate Solutions: Carbon-Reducing Technologies and U.S. Jobs* (Durham, NC: Center for Globalization Governance and Competitiveness, Duke University, November 2008), 15–17.

25. Robert E. Scott, "Unfair China Trade Costs Local Jobs," Economic Policy Institute briefing paper 260, March 23, 2010; Don Lee, "Fighting for 'Made in the USA," *Los Angeles Times*, May 8, 2010.

26. Susan Helper, "Renewing 'Made in the USA,'" *Washington Post*, March 4, 2008.The shift in thinking, since 2008, about the future of American manufacturing in older industrial cities is reflected in the Brookings Institution's Metropolitan Policy Program's "Restoring Prosperity" series. Compare, for example, Jennifer S. Vey's groundbreaking report *Restoring Prosperity: The State Role in Revitalizing America's Older Industrial Cities* (Washington, DC: Brookings Institution, 2007) with Jennifer Bradley, Lavea Brachman, and Bruce Katz, *Restoring Prosperity: Transforming Ohio's Communities for the Next Economy* (Washington, DC: Brookings Institution and Greater Ohio Policy Center, 2010), which argues that we must not divorce innovation from production and calls for stronger federal policy commitment to renewable energy and manufacturing. See also Jennifer S. Vey, John C. Austin, and Jennifer Bradley, *The Next Economy: Economic Recovery and Transformation in the Great Lakes Region* (Washington, DC: Brookings Institution, 2010).

27. For groundbreaking work on business clustering, see Michael Porter, *The Competitive Advantage of Nations* (New York: Free Press, 1990); Paul Krugman, *Geography and Trade* (Cambridge, MA: MIT Press, 1992).

28. Mark Drabenstott, "Past Silos and Smokestacks: Transforming the Rural Economy in the Midwest," Heartland Papers no. 2 (Chicago: Chicago Council on Global Affairs, 2010), 49.

29. Ibid., 8.

30. Mary Ahearn and Jeremy Weber, "Farm Household Economics and Well-Being: Beginning Farmers, Demographics, and Labor Allocations," U.S. Department of Agriculture, Economic Research Service, November 30, 2010.

31. In 2010, Drabenstott was made secretary general of Global Coalition for Efficient Logistics, a worldwide nonprofit based in Geneva, Switzerland. See Rural Policy Research Institute, "RUPRI Announces Senior Leadership Changes," Rupri.org, April 5, 2010 (accessed May 2, 2010).

32. See Center for Innovative Food Technology, http://www.eisc.org/.

33. John Gibney, "ABC World News Recognizes Toledo as 'Solar Valley,'" press release, Regional Growth Partnership, December 18, 2008.

34. Tim Knauss, "Energy Czar Visits Syracuse," *Syracuse Post-Standard,* August 13, 2009.

35. See New York State Foundation for Science, Technology and Innovation, Centers of Excellence, http://www.nystar.state.ny.us/coes.htm (accessed May 21, 2010).

36. See Syracuse Center of Excellence, http://www.syracusecoe.org/coe/images/allmedia/documents/2010/3/CoEprogramWebFINAL.pdf, (accessed May 22, 2010).

37. See "SUNY ESF Biodiesel Program Profiled in NYS Science and Technology Law Center Publication," Green Innovations, Cleantech Center (March 3, 2010).

38. Fitzgerald, *Emerald Cities*, 98–105.

39. Rick Moriarty, "Destiny USA Developer Could Collect $54 Million from State for Removing Dirt," *Syracuse Post-Standard*, April 4, 2010; Rick Moriarty, "Two Syracuse-Area Business Groups, Once Chilly, Plan Merger to Promote Local Economy," *Syracuse Post-Standard*, January 8, 2010.

40. Interview with Robert M. Simpson and Benjamin Walsh, August 24, 2009.

41. Kevin E. Schwab, "Electric Car Manufacturer Selects Syracuse Area," Metropolitan Development Association of Syracuse and CNY, October 23, 2009; "CNY Is a Leader in Race to Produce Electric Vehicles," *MDA Essentials* 1, no. 4 (2009): 1; Tim Knauss, "Electric Car Company Plans to Open Plant in Lysander," *Syracuse Post-Standard*, February 25, 2010. Another manufacturing firm opened in October 2009: Bitzer Scroll, a German firm that makes advanced compressors and other high-efficiency technologies for the refrigeration, air-conditioning, and heat pump industries. It is located at the old GM site in Syracuse and has plans to employ 300 workers. See "Germany-based Bitzer Hosts Grand Opening of CNY Facility," *MDA Essentials* 1, no. 4 (2009): 5. See also Rick Moriarity, "MDA and Syracuse Chamber Announce their New Name, *Syracuse Post-Standard*, May 10, 2010.

42. Rebecca Peterson, "Electric Power Industry Overview 2007," U.S. Energy Information Administration, Independent Statistics and Analysis, n.d.; Louise Guey-Lee, "Renewable Energy Consumption and Electricity Preliminary Statistics 2009," U.S. Energy Information Administration, Independent Statistics and Analysis, August 2010.

43. John M. Broder, "Energy Secretary Serves under a Microscope," *New York Times*, March 22, 2010; Kevin Bullis, "Q&A: Steven Chu," *Technology Review,* April 14, 2009.

44. Office of Electricity Delivery and Energy Reliability, Research and Development Division, U.S. Department of Energy, *Transforming Electricity Delivery: 2007 Strategic Plan* (September 2007), 5. See also Jay Inslee and Bracken Hendricks, *Apollo's Fire* (Washington, DC: Island Press), 303–305.

45. American Society of Civil Engineers, "Report Card for America's Infrastructure: Energy, 2009 Grade D+," n.d.

46. "New Wind Resources Maps and Wind Potential Estimates for the United States," *Wind Powering America,* February 19, 2010.

47. Kevin Bullis, "Costly and Unnecessary New Electricity Grid," *Technology Review*, July 14, 2009. See also David Morris, "A Smart Grid, Yes. A New National Grid, No," *Grist*, March 4, 2009; Matthew L. Wald, "War against a Wind-Rich Super Grid," *New York Times*, April 30, 2010; John Bailey, "East Coast Governors Say National Transmission Grid Limits Local Energy," Newrules.org, June 25, 2009; John Farrell, "Energy Self-Reliant States: Second and Expanded Edition," Newrules.org, October 2009. Arguments for a national transmission superhighway date to 2003. See Department of Energy, Office of Electric of Transmission and Distribution, *Grid 2030: A National Vision for Electricity's Second 100 Years* (Washington, DC: U.S. Department of Energy, 2003). In December 2009, the U.S. Department of Energy awarded $60 million in planning grants (with Recovery Act funding) to explore transmission expansion both within the three regional interconnection networks and among them. Their provisional studies were due in 2011, after this book went to press. See "Secretary Chu Announces Efforts to Strengthen U.S. Electric Transmission Networks," National Renewable Energy Lab, January 4, 2010.

48. See "Distributed Power Generation: Compare Solar Energy with Alternative Energy Sources," Solarbuzz, a Division of the NPD Group, c.2010.

49. Owen, *Green Metropolis*, 264.

50. See California statistics at U.S. Department of Energy Database of State Incentives for Renewables and Efficiency, http://www.dsireusa.org/ (accessed June 2, 2010).

51. Lea Radick, "How a Pioneering University Hopes to Cut Its Carbon Footprint by Half," *New York Times*, May 29, 2009. In the United States, multibuilding district energy systems have come into greater use in recent years, mainly by universities and other large campuses. By early 2011, though, the National Trust for Historic Preservation's Green Lab was piloting several neighborhood-level district energy systems for older, compact commercial districts (which smaller industrial cities have in abundance), where they are most feasible. See Julia Levitt, "Preservation and Sustainability: The District Approach," September 22, 2010, http://www.metropolismag.com/pov/20100922/preservation-and-sustainability-the-district-approach (accessed March 21, 2011). It should also be noted that "big-g" geothermal, which has received big federal funding, is associated with earthquake risks. See James Glanz, "Geothermal Drilling Safeguards Imposed," *New York Times*, January 15, 2010.

52. Franny Ritchie, "Missouri Town Goes Off the Grid," *Planetizen*, August 4, 2008; Jane Burgermeister, "Renewable Energy Powers Italian Town and Its Economy," Renewableenergyworld.com, December 17, 2007 (accessed April 15, 2010).

53. See "C40 Cities: An Introduction," Clinton Climate Initiative, Clinton Foundation, c.2010, http://www.c40cities.org(accessed January 6, 2011). See also the work of ICLEI: Local Governments for Sustainability, whose U.S.group made available a sustainability planning toolkit suited to cities of all sizes in 2009: http://www.icleiusa.org/action-center/planning/sustainability-planning-toolkit (accessed March 11, 2011).

54. U.S. Energy Information Administration, Independent Statistics and Analysis, "Electric Power Industry Overview 2007," http://www.eia.doe.gov/electricity/page/prim2/toc2.html (accessed May 15, 2010); National Renewable Energy Lab, "NREL Highlights Utility Green Power Leaders," May 3, 2010, http://www.nrel.gov/news/press/2010/838.html (accessed May 15, 2010).

55. John Randolph and Gilbert M. Masters, *Energy for Sustainability: Technology, Planning, Policy* (Washington, DC: Island Press, 2008); John Randolph, *Environmental Land Use, Planning, and Management,* 2nd ed. (Washington, DC: Island Press, 2011). My thanks go to Randolph for sharing new material for the second edition on localized municipal energy arrangements. For a less sanguine view of creating new municipal utilities, which stresses that their up-front capital costs would minimize price differences with services offered by IOUs, and voices political concerns about local control, see Massachusetts Department of Energy Resources, *Municipal Utility Study: Technical Report,* January 28, 2010.

56. U.S. Department of Energy Database of State Incentives for Renewables and Efficiency, Texas: Incentives/Policies for Renewables and Efficiency, Summary, http://www.dsireusa.org/incentives/incentive.cfm?Incentive_Code=TX11R&re=1&ee= (accessed January 4, 2011); Lisa A. Lemons, "Denton Municipal Electric Secures Green Energy," Denton Municipal Electric and City of Denton, April 22, 2009, http://www.dentonedp.com/news_case_studies/pdf/PR_0422_RenewableEnergyPressConference.pdf (accessed May 28, 2010); Mary Logan Barmeyer, "Energy/Denton, Texas," Smartercities.nrdc.org, June 30, 2010; D. C. Denison, "Holyoke Chosen for Computing Center," *Boston Globe*, June 10, 2009; Jaclyn C. Stevenson, "Power Trip: Holyoke G&E Raises the Bar for Municipal Utilities," *Business West, AllBusiness, A D&B Company*, February 1, 2005. Paul Gipe, "Gainesville Muni Reaches 4 MW in Solar Reservations," Wind-Works.org, February 27, 2009; Kate Galbraith, "Europe's Way of Encouraging Solar Power Arrives in the U.S.," *New York Times*, March 12, 200. For a brief overview of feed-in tariffs, see Fitzgerald, *Emerald Cities*, 6.

57. See "Los Angeles Should Adopt Solar Feed-In Tariffs," RenewableEnergyFocus.com, April 9, 2010; Dan Haugen, "Why Isn't the U.S. Embracing Feed-in Tariffs?" *SolveClimate News*, March 24, 2009.

58. Adrian Moore, "Utilities Takeover by Cities a Mistake," *Reason*, January 14, 2003.

59. Northeast Ohio Public Energy Council, "First Energy Solutions and NOPEC Enter into Nine-Year Agreement," Nopecinfo.org, December 2, 2009; FirstEnergy, *2009 Annual Report*, 2009, http://www.firstenergycorp.com/financialreports/files/2009AnnualReport.pdf (accessed May 29, 2010).

60. Deborah Kahn "Community Bids to Bypass Utilities Facing Hurdles in Calif.," *New York Times*, January 7, 2010; Tam Hunt, "Why Is Community Choice Aggregation So Promising?" RenewableEnergyWorld.com, August 20, 2008.

61. See, for example, Chicago's 2001 agreement with IOU Commonwealth Edison to purchase 20 percent of its electricity from renewable sources beginning in 2006. "Chicago—Green Power Purchasing," DSIRE, March 22, 2010. In May 2010, the city allowed another private utility, BlueStar Energy, to compete for residential

business offering 100 percent renewable-based electricity services. See "BlueStar Energy Offers IL Residents Alternative to ComEd's Energy Tax Proposal," PR Newswire, United Business Media, May 5, 2010.

62. E-mail exchange with Cleantech America CEO Bill Barnes, October 3, 2008. At the time, his company was building a utility-scale solar farm in the Fresno, California, area.

63. For an excellent overview of biofuel issues, see Inslee and Hendricks, *Apollo's Fire*, 147–177.

64. Jim Lane, "Earth Day Stunner: POET Sets 2022 Cellulosic Ethanol Target of 3.5 Billion Gallons Per Year," *Biofuels Digest*, April 21, 2010.

65. Inslee and Hendricks, *Apollo's Fire*, 158–159.

66. See "Crave Brothers Farm Reaps Big Benefits with Energy Efficiency and Renewable Energy," Wisconsin Focus on Energy, 2009.

67. Brian E. Clark, "Anaerobic Digesters Turning State's Dairy Waste into Power," Wisbusiness.com, April 14, 2008.

68. U.S. Environmental Protection Agency "Anerobic Digestion," AgSTAR, http://www.epa.gov/agstar/anaerobic/index.html (accessed March 11 , 2011).

69. Cronon, *Nature's Metropolis*, 55–77; Lawrence A. Baker, *Water Environment of Cities* (New York: Springer, 2009), 1–9.

70. Julie Smith-Glavin, "Enel Announces Improvements to Be Made to Historic Great Stone Dam Beginning Summer 2007," Enel North America, June18, 2007. Most of this discussion is indebted to an interview with Russ Cohen, August 6, 2008.

71. See William Arthur Atkins, "Hydroelectric Power," in *Water Encyclopedia* (n.d.); U.S. Department of the Interior, "Interior, Energy, and Army Corps of Engineers Sign Memorandum of Understanding on Hydropower," March 24, 2010.

72. See "Overview: Green Building," Open Square, c.2010, http://opensquare.com/green-energy.php (accessed May 24, 2010).

73. Union of Concerned Scientists, "How Hydrokinetic Energy Works," www.ucsusa.org, April 28, 2008; Pew Center on Global Climate Change, "Hydrokinetic Electric Power Generation," December 2009.

74. Douglas G. Hall et al., *Water Energy Resources of the United States with Emphasis on Low Head/Low Power Resources* (Idaho Falls, ID: National Engineering and Environmental Laboratory, April 2004).

75. See "Hastings Hydrokinetic Power Station, USA," www.power-technology.com, n.d., http://www.power-technology.com/projects/hastingshydrokinetic/ (accessed May 24, 2010).

76. See Free Flow Power: http://www.free-flow-power.com/Technology.html.

77. Federal Energy Regulatory Commission, "Hydrokinetic Projects," ferc.gov, May 18, 2010; Jacques Beaudry-Losique, "Statement of Jacques Beaudry-Losique, Deputy Assistant Secretary for Renewable Energy, Office of Energy Efficiency and Renewable Energy, U.S. Department of Energy, before the Committee on Science and Technology, Subcommittee on Energy and Environment, U.S. House of

Representatives, Hearing Examining Marine and Hydrokinetic Energy Technology: Finding the Path to Commercialization," December 3, 2009.

78. See, for example, Efrain Viscarolasaga, "Going with the Free Flow of River Power Using Submerged Turbines," Masshightech.com, September 5, 2008.

79. Juliet Eilperin and Anne E. Komblut, "President Obama Opens New Areas to Offshore Drilling," *Washington Post*, April 1, 2010; "Marcellus Shale–Appalachian Basin Natural Gas Play: New Research Results Surprise Everyone on the Potential of This Well-Known Devonian Black Shale," geology.com, 2010, \ http://geology.com/articles/marcellus-shale.shtml (accessed May 25, 2010); Dan O'Brien, "V&M Exec: 'My Knees Are Shaking!'" Business-journal.com, February 16, 2010.

Chapter 6

1. Telephone interview with Robert Forrant, June 4, 2010. See Robert Forrant, *Metal Fatigue: American Bosch and Demise of Metalworking in the Connecticut River Valley* (Amityville, NY: Baywood, 2009).

2. See the Web sites for the Regional Technology Corporation, http://www.rtccentral.com/about.php, and the Western Massachusetts Economic Development Corporation, http://www.westernmassedc.com/. Their strategy is based on a Battelle Consultants report: Battelle Institute Technology Partnership Practice, *Western Massachusetts Regional Technology Audit and Network Identification Assessment: Building Technology Networks for Regional Economic Competitiveness* (January 2001).

3. Several people I interviewed for this book also complained about the absence of high-tech manufacturing skills in today's labor force. See Motoko Rich, "Factory Jobs Return, But Employers Find Skills Shortage," *New York Times*, July 1, 2010.

4. Benjamin Forman et al., *Building for the Future: Foundations for a Springfield Comprehensive Growth Strategy* (Boston: MassINC, June 2009), 41.

5. Jane Jacobs, *Cities and the Wealth of Nations: Principles of Economic Life* (New York: Vintage Books, 1984), 35–47. For her discussion of import substitution as it played out in Uruguay, see 60–64.

6. Ibid., 38.

7. Jacobs, *The Economy of Cities* (New York: Vintage Books, 1969), 163, and *Cities and the Wealth of Nations*, 39–41.

8. Richard Florida, *The Great Reset* (New York: HarperCollins, 2010), 46. See also Pierre Desrochers and Gert-Jan Hospers, "Cities and the Economic Development of Nations: An Essay on Jane Jacobs's Contribution to Economic Theory," *Canadian Journal of Regional Science* 30, no. 1 (2007): 115–130.

9. Eswar Presad, "The U.S.-China Economic Relationship: Shifts and Twists in the Balance of Power," testimony before the Economic and Security Review Commission, Brookings.edu, February 25, 2010, updated March 10, 2010; David Barboza, "As China's Wages Rise, Export Prices Could Follow," *New York Times*, June 7, 2010.

10. Jacobs, *Cities and the Wealth of Nations*, 71.

11. Ibid., 39–41.

12. "About the New Rules Project," www.newrules.org.

13. The full complement of localist strategies—and the arguments of their critics—is nicely summarized and placed in historical context by David J. Hess, *Localist Movements in a Global Economy* (Cambridge, MA: MIT Press, 2009). For the most comprehensive advocacy studies of independent business localism, see Stacy Mitchell, *Big-Box Swindle: The True Cost of Mega-Retailers and the Fight for America's Independent Businesses* (Boston: Beacon Press, 2006); Michael Shuman, *The Small-Mart Revolution: How Local Businesses Are Beating the Global Competition* (San Francisco: Berrett-Koehler, 2006), and *Going Local: Creating Self-Reliant Communities in a Global Age* (New York: Routledge, 2000); David Morris and Karl Hess, *Neighborhood Power: The New Localism* (Boston: Beacon Press, 1975). See also the Web site of the Institute for Local Self-Reliance, founded by David Morris: ilsr.org. On reforming global trade policies, see Thad Williamson, David Imbroscio, and Gar Alperowitz, *Making a Place for Community: Local Democracy in a Global Era* (New York: Routledge, 2002), and Colin Hines, *Localization: A Global Manifesto* (London: Earthscan, 2000). For corporate-liberal arguments against the limits of globalization, with suggestions such as putting a legal cap on the national trade deficit and offering tax incentives to companies that invest in training domestic workers, see Ralph Gormory and William Baumol, *Global Trade and Conflicting National Interests* (Cambridge, MA: MIT Press, 2001).

14. E. F. Schumacher, *Small Is Beautiful: Economics as if People Mattered* (New York: Harper Colophon, 1973); David J. Morris and Karl Hess, *Neighborhood Power: The New Localism* (Boston: Beacon Press, 1975).

15. Hess, *Localist Movements,* 113–133. The statistic comes from Stacy Mitchell, "Neighborhood Stores: An Overlooked Strategy for Fighting Global Warming," *Grist*, August 19, 2009.

16. Hess, *Localist Movements*, 84–87, 107.

17. Telephone interview with Stacy Mitchell, October 26, 2010. See also Mitchell, "Neighborhood Stores."

18. Mitchell, *Big-Box Swindle*, 45.

19. Dan Houston, "Local Works: Examining the Impact of Independent Business on the West Michigan Economy," Civiceconomics.com (September 2008).
21. See also David Neumark, Junfu Zhang, and Stephen Ciccarella, "The Effects of Wal-Mart on Local Labor Markets," IZA discussion paper no. 2545, January 2007; T. Arindrajit Dube, William Lester, and Barry Eidlin, "A Downward Push: The Impact of Wal-Mart Stores on Retail Wages and Benefits" (Berkeley: University of California, Berkeley, Center for Labor Research and Education, December 2007), which, among other findings, shows that nationally, in 2000, "total earnings of retail workers nationwide were reduced by $4.5 billion due to Wal-Mart's presence," and that most of these losses were concentrated in metropolitan areas since three-quarters of the stores Wal-Mart built in the 1990s were in metropolitan counties; and Jared Bernstein, Josh Bivens, and Arindrajit Dube, "Wrestling with

Wal-Mart: Tradeoffs between Prices, Profits, and Wages," EPI working paper 276, June 14, 2006.

20. On the act's provisions and implementation procedures, see www.informedgrowthact.com. On the 2010 amendment to the act, see http://www.mainemerchants.org/legislative-affairs/position.php?lID=291.

21. Shuman, *Going Local,* 99.

22. National Commission on Excellence in Education, *A Nation at Risk: The Imperative for Educational Reform,* (Washington, DC: U.S. GPO, 1983); Diane Ravitch, *The Death and Life of the Great American School System: How Testing and Choice Are Undermining Education* (New York: Basic Books, 2010).

23. Gerald Grant, *Hope and Despair in the American City: Why There Are No Bad Schools in Raleigh* (Cambridge, MA: Harvard University Press, 2009).

24. James Coleman et al., *Equality of Educational Opportunity* (Washington, DC: Government Printing Office, 1966). This massive study mandated by the Civil Rights Act of 1964, which surveyed some 600,000 students and teachers, was a lightning rod for controversy, in part because it led to the busing programs of the early 1970s. In 1975, Coleman disavowed busing as a policy that fostered white flight as long as suburban districts were separated from urban ones. It also collided with African American backlash against the integrationist aims of the civil rights movement. For sociological studies that buttress the Coleman Report's claims, on which Grant relies and which shaped Wake County's approach, see William Julius Wilson, *The Declining Significance of Race: Blacks and Changing American Institutions* (Chicago: University of Chicago Press,1980), and *The Truly Disadvantaged: The Inner City, Underclass and Public Policy* (Chicago: University of Chicago Press, 1990); Russell Rumberger and Gregory Palardy, "Does Segregation Still Matter? The Impact of Student Composition of Academic Achievement in High School," *Teachers College Record* 107 (2005): 1999–2045.

25. Grant, *Hope and Despair,* 88.

26. Grant, *Hope and Despair*, 104, 120–128; Wake County Public School System Evaluation and Research Department, "2008-2009 WCPSS Dropout Rate, E&R Report No. 10.03," March 2010, http://www.wcpss.net/evaluation-research/reports/2010/1003dropout08_09.pdf (accessed March 21, 2011).

27. On La Crosse, see Richard D. Kahlenberg, *Rescuing Brown* v. *Board of Education: Profiles of Twelve School Districts Pursuing Socioeconomic Integration* (New York: Century Foundation, 2009); Grant, *Hope and Despair*, 163–164; David Eichenthal and Tracy Windeknecht, "Chattanooga, Tennessee: A Restoring Prosperity Case Study" (Washington, DC: Metropolitan Policy Program, Brookings Institution, September 2008).

28. On Hartford, see Richard Kahlenberg, "Can School Integration Work in the Cities?" Taking Note blog, July 24, 2008, http://takingnote.tcf.org/2008/07/can-school-inte.html, (accessed June 12, 2010), and Robert A. Frahm, "School Choice: 'The Most Efficient' Way to Desegregate," *Connecticut Mirror*, February 2, 2010. Payzant quoted in Emily Bazelon, "The Next Kind of Integration," *New York Times Magazine,* July 20, 2008, an excellent summary of the legal, scholarly,

and political context within which school districts have been scrambling in the aftermath of the June 28, 2007 Louisville and Seattle Supreme Court decisions in *Parents Involved in Community Schools Inc. v. Seattle School District* and *Meredith v. Jefferson County (Ky.) Board of Education.*

29. Sean Price, "Charter Schools: Resegregating America?" Teachingtolerance.org blog, February 5, 2010, http://www.tolerance.org/blog/charter-schools-resegregating-america (accessed June 10, 2010); Gary Orfield and Chungmei Lee, *Historic Reversals, Accelerating Resegregation, and the Need for New Integration Strategies* (Los Angeles: UCLA, 2007); Amanda Paulson, "Resegregation of U.S. Schools Deepening," *Christian Science Monitor*, January 25, 2008.

30. T. Keung Hiu, "Critics Say Survey Tells Wake Board to Rethink," *Raleigh News and Observer*, February 5, 2010; Chris Fitzsimon, "Missteps in the March to Resegregation," NCPolicyWatch.com, February 3, 2010; Stephanie McCrummen, "Republican School Board in N.C. Backed by Tea Party Abolishes Integration Policy," *Washington Post*, January 12, 2011; Chris Kromm, "Pope the Vote? Conservative N.C. Benefactor Steps Up Funding to Nonprofits for Ads During Election Season, *Facing South*, October 8, 2010, http://www.southernstudies.org/2010/10/pope-the-vote-conservative-nc-benefactor-steps-up-funding-to-nonprofits-for-ads-during-election-season.html (accessed March 21, 2011). See also "John Tedesco Wake County School Board at Tea Party, April 15, 2010," http://www.youtube.com/watch?v=r9rXleSIWdM.

31. "John Tedesco: Candidate for Wake County Board of Education District 2," Indyweek.com, May 12, 2009, http://www.indyweek.com/indyweek/john-tedesco/Content?oid=1217511 (accessed June 11, 2010).

32. T. Keung Hui, "Two Businessmen Invested Big in Schools Race," *Raleigh News and Observer*, February 9, 2010; Rob Schofield, "A History Lesson for Our Public Schools," NCPolicyWatch.com, February 13, 2010.

33. Kate Zernike and Megan Thee-Brenan, "Poll Finds Tea Party Backers Wealthier and More Educated," *New York Times*, April 14, 2010.

34. In 1965, the ratio between the poorest and the wealthiest American citizens stood at 25 to 1; by 2005, it was 625 to 1. Between 2000 and 2007, median household income (around $46,000 a year) had dropped by $1,175, adjusted for inflation, while the average family of four spent $4,655 more on basic expenses, such as gas, housing, and health insurance. None of these figures takes into account the soaring costs of higher education, a standard marker of middle-class status. Since the early 1970s, when one working parent was the middle-class norm, discretionary income has plummeted by more than 11 percent, even with both parents now working. In other words, the American middle classes are leading a far more precarious existence, one health care crisis or layoff away from penury. See Andrew Sum, Ishwar Khatiwada, and Sheila Palma, "Labor Underutilization Problems of U.S. Workers across Household Income Groups at the End of the Great Recession: A Truly Great Depression among the Nation's Low Income Workers amidst Full Employment among the Most Affluent" (Boston: Center for Labor Market Studies, Northeastern University, February 2010); Elizabeth Warren, "The Vanishing Middle Class," in *Ending Poverty in America: How to Restore*

the American Dream, ed. John Edwards, Marion Crain, and Arne L. Kalleberg (New York: New Press, 2007), 38–52.

35. Thomas Frank, *What's the Matter with Kansas?* (New York: Metropolitan Books, 2004).

36. C. Wright Mills and Melville. J. Ulmer; directed by John M. Blair, *Small Business and Civic Welfare*, Report for the Smaller War Plants Corporation, U.S. War Production Board, January 1946. My deepest thanks belong to John Summers for bringing this report to my attention, loaning me his copy, and sharing the results of his archival labors. His sleuthing revealed that the six cities Mills studied were Flint, Kalamazoo, Grand Rapids, and Pontiac, in Michigan; Nashua, New Hampshire; and Rome, New York.

37. Nancy C. Unger, *Fighting Bob La Follette: The Righteous Reformer* (Chapel Hill: University of North Carolina Press, 2000); David P. Thelan, *Robert M. La Follette and the Insurgent Spirit* (Madison: University of Wisconsin Press, 1985). On the anti–chain store movement see Hess, *Localist Movements*, 114; Bethany Moreton, *To Serve God and Wal-Mart: The Making of Christian Enterprise* (Cambridge, MA: Harvard University Press, 2009); Daniel Scroop, "The Anti–Chain Store Movement and the Politics of Consumption," *American Quarterly* 60 (2008): 925–949; and Richard Schragger, "The Anti–Chain Store Movement, Localist Ideology, and the Remnants of the Progressive Constitution, 1920–1940," *Iowa Law Review* 90 (2005): 1011–1094. The anti–chain store movement was most vocal in the South and Midwest—strongholds of populist insurgency a generation before. Unlike today's localists, who seek to level the playing field by removing tax breaks given to big-retail developers (a practice that began with chains on the federal level in the 1950s), they opted to tax the big-box equivalents of their day, with the intent of forcing them to raise prices. By 1936, more than 800 "chain taxes" had been passed by states and localities, and in 1938 Congress introduced a "community preservation" bill that would have federalized these measures. At that point, however, so many labor, consumer, and commercial farming interests were hooked into the big-retail distribution model that the well-organized retail lobby was able to kill the bill. Moreton, *To Serve God and Wal-Mart*, 18–19.

38. Louis D. Brandeis, *Other People's Money and How the Bankers Use It* (Washington, DC: National Home Library Foundation, 1933); Melvin I. Urofsky, *Louis D. Brandeis: A Life* (New York: Pantheon, 2010).

39. Mumford, *City in History*, 536–537.

40. Richard Hofstadter, *The Age of Reform* (New York: Vintage, 1955), 243–244, 255.

Selected Bibliography

Alexander, Christopher. "A City Is Not a Tree." *Architectural Forum* 122 (April 1965): 58–62; (May 1965): 58–61.

Alexander, Christopher, with Sara Ishikawa and Murray Silverstein. *A Pattern Language: Towns, Buildings, and Construction*. New York: Oxford University Press, 1977.

American Society of Civil Engineers. *Report Card for America's Infrastructure: Energy, 2009 Grade D+*. Reston, VA: American Society of Civil Engineers, 2009.

Arendt, Randall. *Conservation Design for Subdivisions: Guide to Creating Open Space Networks*. Washington, DC: Island Press, 1996.

Arendt, Randall. *Growing Greener: Putting Conservation into Local Plans and Ordinances*. Washington, DC: Island Press, 1999.

Ayee, Gloria, Marcy Lowe, and Gary Gereffi. *Manufacturing Climate Solutions: Carbon-Reducing Technologies and U.S. Jobs*. Durham, NC: Duke University, 2008.

Bailey, John. "East Coast Governors Say National Transmission Grid Limits Local Energy." Newrules.org, June 25, 2009.

Baker, Lawrence A. *Water Environment of Cities*. New York: Springer, 2009.

Barboza, David. "As China's Wages Rise, Export Prices Could Follow." *New York Times*, June 7, 2010.

Barmeyer, Mary Logan. "Energy/Denton, Texas." Smartercities.nrdc.org, June 30, 2010.

Battelle Institute Technology Partnership Practice. *Western Massachusetts Regional Technology Audit and Network Identification Assessment: Building Technology Networks for Regional Economic Competitiveness*. Columbus, OH: Battelle Institute, January 2001.

Baue, Bill. "A Hybrid Model: Co-op and Nonprofits Launch Energy-Efficiency Company." *Sea Change*, radio broadcast, January 20, 2010.

Bazelon, Emily. "The Next Kind of Integration." *New York Times*, July 20, 2008.

Beatley, Timothy. *Green Urbanism: Learning from European Cities*. Washington, DC: Island Press, 2000.

Bell, Daniel. *The Coming of Post-Industrial Society: A Venture in Social Forecasting*. New York: Harper Colophon, 1974.

Bell, David, and Mark Jayne, eds. *Small Cities: Urban Experience beyond the Metropolis*. London: Routledge, 2006.

Bell, David, and Mark Jayne. "Small Cities? Towards a Research Agenda." *International Journal of Urban and Regional Research* 33 (2009): 683–699.

Benfield, Kaid. "They Are Stardust. They Are Golden. But Are They Right about 'Shrinking Cities'?" Switchboard.nrdc.org blog, July 2, 2009.

Beniston, Ian. Memo to Community and Neighborhood Leaders on Strategic Demolition Policy. Mahoning Valley Organizing Committee. Mvorganizing.org., January 20, 2009. http://www.mvorganizing.org/downloads/MVOC_Strategic_Demolition_Policy_Brief.pdf. Accessed March 10, 2010.

Berg, Nate. "Brave New Codes." *Architect,* July 2010. http://www.architectmagazine.com/codes-and-standards/brave-new-codes.aspx. Accessed March 18, 2011.

Berg, Nate. "Redressing Strip Malls." Planetizen blog, November 14, 2008, http://www.planetizen.com/node/36042. Accessed February 25, 2010.

Bernstein, Jared, Josh Bivens, and Arindrajit Dube. "Wrestling with Wal-Mart: Tradeoffs between Prices, Profits, and Wages." EPI working paper 276. Washington, DC: Economic Policy Institute, 2006.

Berry, Brian. *Geography of Market Centers and Retail Distribution*. Englewood Cliffs, NJ: Prentice Hall, 1967.

Berry, Brian J. L., and Chauncy D. Harris. "Walter Christaller: An Appreciation." *Geographical Review* 60 (1970): 1.

Berry, Wendell. *The Unsettling of America: Culture and Agriculture*. 3rd ed. Berkeley, CA: Sierra Club Books, 1997.

Bird, Lori, Caroline Chapman, Jeff Logan, Jenny Sumner, and Walter Short. *Evaluating Renewable Portfolio Standards and Carbon Cap Scenarios in the U.S. Electric Sector.* Technical report NREL/TP-6A2–48258I. Golden, CO: U.S. Department of Energy, 2010.

Black, Russell Van Nest. *Planning for the Small American City*. Chicago: Public Administration Service, 1934.

Bluestone, Barry, and Bennett Harrison. *The Deindustrialization of America: Plant Closings, Community Abandonment, and the Dismantling of Basic Industry*. New York: Basic Books, 1982.

Bohl, Charles C., and Elizabeth Plater-Zyberk. "Building Community across the Rural-to-Urban Transect." *Places* 18, no. 1 (2006): 4–17.

Bookchin, Murray. *The Limits of the City*. New York: Harper & Row, 1974.

Bosworth, Kimberly. "Conservation Subdivision Design: Perceptions and Reality." Master's thesis, University of Michigan, 2007.

Bradley, Jennifer, Lavea Brachman, and Bruce Katz. *Restoring Prosperity: Transforming Ohio's Communities for the Next Economy*. Washington, DC: Brookings Institution and Greater Ohio Policy Center, 2010.

Bradsher, Keith. "China Sees Growth Engine in a Web of Fast Trains." *New York Times*, February 12, 2010.

Bralich, John. "Developing Methods to Establish an Urban Wetland Mitigation Bank on Youngstown's East Side." Youngstown, OH: Youngstown State University Center for Urban and Regional Studies, 2009.

Briggs, Suzanne, et al. "Real Food, Real Choices: Connecting SNAP Recipients with Farmers Markets." San Francisco: Tides Foundation, June 2010.

Brockman, Henry. 2001. *Organic Matters*. Congerville, IL: TerraBooks.

Broder, John M. "Energy Secretary Serves under a Microscope." *New York Times*, March 22, 2010.

Bryce, Herrington J. *Planning Smaller Cities*. Lexington, MA: D. C. Heath, 1979.

Bryce, Herrington J., ed. *Small Cities in Transition: The Dynamics of Growth and Decline*. Cambridge, MA: Ballinger, 1977.

Buenker, John D., and Theodore Mesmer. "A Separate Universe? An Exploratory Effort at Defining the Small City." *Indiana Magazine of History* 99 (2003): 331–352.

Bullis, Kevin. "Costly and Unnecessary New Electricity Grid." *Technology Review* (July 14, 2009) http://www.technologyreview.com/energy/22997/?a=f. Accessed October 15, 2010.

Bullis, Kevin. "Q&A: Steven Chu." *Technology Review* April 14, 2009 http://www.technologyreview.com/business/22651/. Accessed October 14, 2010.

Bullis, Kevin. "U.S. Solar Market to Double in the Next Year." *Technology Review* (February 8, 2010) http://www.technologyreview.com/energy/24498/. Accessed October 15, 2010.

Burayidi, M., ed. *Downtowns: Revitalizing the Centers of Small Urban Communities*. New York: Routledge, 2001.

Burnham-Jones, Ellen, and June Williamson. *Retrofitting Suburbia: Urban Design Solutions for Redesigning Suburbs*. Hoboken, NJ: Wiley, 2008.

Cahn, Naomi, and June Carbone. *Red Families v. Blue Families: Legal Polarization and the Creation of Culture*. New York: Oxford University Press, 2010.

Caldwell, Emily. "Ohio Church Serves God, Community with Small-Scale Urban Gardening." Farmanddairy.com, July 9, 2009.

Calthorpe, Peter, and William Fulton. *The Regional City: Planning for the End of Sprawl*. Washington, DC: Island Press, 2001.

Carpenter, Novella. *Farm City: The Education of an Urban Farmer*. New York: Penguin, 2009.

Charles Stewart Mott Foundation. *"Flint: A Snapshot in Time": Charles Stewart Mott Foundation 2008 Annual Report*. Flint, MI: Charles Stewart Mott Foundation, 2008.

Christaller, Walter. *Central Places in Southern Germany*. Translated by Carlisle W. Baksin. Englewood Cliffs, NJ: Prentice Hall, 1966.

Christensen, Julia. *Big Box Reuse*. Cambridge, MA: MIT Press, 2008.

Clancy Kate, Janet Hammer, and Debra Lippoldt. "Food Policy Councils: Past, Present, and Future." In *Remaking the North American Food System*. Edited by C. Clare Hinrichs and Thomas A. Lyson. Lincoln, Nebraska: University of Nebraska Press, 2007, 121–140.

Clark, Brian E. "Anaerobic Digesters Turning State's Dairy Waste into Power." Wisbusiness.com, April 14, 2008.

"CNY Is a Leader in Race to Produce Electric Vehicles." *MDA Essentials* 1 (2009): 4.

Coleman, James, Ernest Q. Campbell, Carol J. Hobson, James McPartland, Alexander M. Mood, Frederic D. Weinfeld, and Robert L. York. *Equality of Educational Opportunity*. Washington, DC: Government Printing Office, 1966.

Congress for the New Urbanism. "Freeways without Futures." Cnu.org, 2008.

Conner, David S., William A. Knudson, Michael W. Hamm, and H. Christopher Peterson. "The Food System as an Economic Driver: Strategies and Applications for Michigan." *Journal of Hunger and Environmental Nutrition* 3 (2008): 371–383.

Connolly, James J. "Decentering Urban History: Peripheral Cities in the Modern World." *Journal of Urban History* 35 (2008): 3–14.

Connolly, James J., ed. *After the Factory: Reinventing America's Industrial Small Cities*. Lanham, MD: Rowman and Littlefield, 2010.

Cronon, William. *Nature's Metropolis: Chicago and the Great West*. New York: Norton, 1991.

Davis, Lisa Selin. "The Secret Mall Gardens of Cleveland." Grist.org, March 17, 2010.

Denison, D. C. "Holyoke Chosen for Computing Center." *Boston Globe*, June 10, 2009.

DeParle, Jason, and Robert Gebeloff. "Food Stamp Use Soars, and Stigma Fades." *New York Times*. November 29, 2009, A-1.

Department of Energy. Office of Electric of Transmission and Distribution. *Grid 2030: A National Vision for Electricity's Second 100 Years. U.S. Department of Energy*. Washington, DC: Government Printing Office, 2003.

Desrochers, Pierre, and Gert-Jan Hospers. "Cities and the Economic Development of Nations: An Essay on Jane Jacobs's Contribution to Economic Theory." *Canadian Journal of Regional Science* 30 (2007): 115–130.

Dowie, Mark. *American Foundations*. Cambridge, MA: MIT Press, 2001.

Dowie, Mark. "Food among the Ruins." Guernicamag.com, August 2009.

Drabenstott, Mark. "Past Silos and Smokestacks: Transforming the Rural Economy in the Midwest." Chicago: Chicago Council on Global Affairs, 2010.

Drucker, Peter. *The Effective Executive*. New York: Harper & Row, 1966.

Duany Plater-Zyberk & Company, LLC. "Agricultural Urbanism." Draft report, March 25, 2009.

Duany, Andrés. "Introduction to the Special Issue Dedicated to the Transect." *Journal of Urban Design* 7 (2002): 251–260.

Duany, Andrés, Elizabeth Plater-Zyberk, and Jeff Speck. *Suburban Nation: The Rise of Sprawl and the Decline of the American Dream.* New York: North Point Press, 2001.

Duany, Andrés and Emily Talen. "Making the Good Easy: The Smart Code Alternative." *Fordham Urban Law Journal* 29 (2002): 1445–1468.

Duany, Andrés, Jeff Speck, and Mike Lydon, *The Smart Growth Manual.* New York: McGraw-Hill, 2009.

Dubb, Steve. Interview with Dan Kildee. Community-Wealth.org, June 2009.

Dube, T. Arindrajit, William Lester, and Barry Eidlin, *A Downward Push: The Impact of Wal-Mart Stores on Retail Wages and Benefits.* Berkeley: University of California, December 2007.

Dunham-Jones, Ellen, and June Williamson. *Retrofitting Suburbia: Urban Design Solutions for Redesigning Suburbs.* Hoboken, NJ: Wiley, 2008.

Ehrenhalt, Alan. "Trading Places: The Demographic Inversion of the American City." *New Republic,* August 13, 2008.

Eichenthal, David, and Tracy Windeknecht. *Chattanooga, Tennessee: A Restoring Prosperity Case Study.* Washington, DC: Brookings Institution Metropolitan Policy Program, 2008.

Eilperin, Juliet, and Anne E. Komblut. "President Obama Opens New Areas to Offshore Drilling." *Washington Post,* April 1, 2010.

Emerine, Dan, Christine Shenot, Mary Kay Bailey, Lee Sobel, and Megan Susman, *This Is Smart Growth.* Washington, DC: Smart Growth Network, 2006.

Engardio, Pete. "In Detroit, Is There Life after the Big 3?" *New York Times,* February 14, 2010.

Erickcek, George A., and Hannah McKinney. "'Small Cities Blues': Looking for Growth Factors in Small and Medium-Sized Cities." Working paper 04-100. Kalamazoo, MI: Upjohn Institute for Employment Research, June 2004.

Ewing, Reid. "Is Los Angeles–Style Sprawl Desirable?" *Journal of the American Planning Association* 63 (1997): 107–126.

Ewing, Reid, Rolf Pendell, and Don Chen, *Measuring Sprawl and Its Impact.* Washington, DC: Smart Growth America, 2002.

Farrell, John. "Energy Self-Reliant States: Second and Expanded Edition." Newrules.org, October 2009.

Favro, Tony. "American Cities Seek to Discover Their Right Size." Citymayors.com, April 5, 2010.

Fishman, Robert. *Bourgeois Utopias: The Rise and Fall of Suburbia.* New York: Basic Books, 1987.

Fitzgerald, Joan. *Emerald Cities: Urban Sustainability and Economic Development.* New York: Oxford University Press, 2010.

Fitzsimon, Chris. "Missteps in the March to Resegregation." NCPolicyWatch.com, February 3, 2010.

Flint, Anthony. *This Land: The Battle over Sprawl and the Future of America.* Baltimore, MD: Johns Hopkins University Press, 2006.

Flint, Anthony. *Wrestling with Moses: How Jane Jacobs Took on New York's Master Builder and Transformed the American City*. New York: Random House, 2009.

Flint Futures Group. *Reimagining Chevy in the Hole*. Ann Arbor: School of Natural Resources and Environment, University of Michigan, 2007.

Florida, Richard. *The Flight of the Creative Class: The New Global Competition for Talent*. New York: HarperBusiness, 2005.

Florida, Richard. *The Great Reset: How New Ways of Living and Working Drive Post-Crash Prosperity*. New York: HarperCollins, 2010.

Florida, Richard. "How the Crash Will Reshape America." *Atlantic Monthly* (March 2009): 23–36.

Florida, Richard. *The Rise of the Creative Class and How It's Transforming Work, Leisure, Community, and Everyday Life*. New York: Basic Books, 2002.

Florida, Richard. *Who's Your City? How the Creative Economy Is Making Where You Live the Most Important Decision of Your Life*. New York: Basic, 2008.

Forman, Benjamin, Ed Lambert, John Schneider, Dana Ansel, and Jason Silva. *Building for the Future: Foundations for a Springfield Comprehensive Growth Strategy*. Boston: MassINC, 2009.

Forman, Minehanna. "Race Dynamic Seen as Obstacle in Detroit Urban Farming." *Michigan Messenger*, October 30, 2009.

Forman, Richard T. T. *Urban Regions: Ecology and Planning Beyond the City*. Cambridge: Cambridge University Press, 2008.

Forrant, Robert. *Metal Fatigue: American Bosch and the Demise of Metalworking in the Connecticut River Valley*. Amityville, NY: Baywood, 2009.

Fox, Radhika, and Miriam Axel-Lute. *To Be Strong Again: Renewing the Promise in Smaller Industrial Cities*. New York: PolicyLink, 2008.

Fox, Richard Wightman. "Epitaph for Middletown: Robert S. Lynd and the Analysis of Consumer Culture." In *The Culture of Consumption: Critical Essays in American History, 1880–198*. Edited by Richard Wrightman Fox and T. J. Jackson Lears. New York: Pantheon, 1983.

Frahm, Robert A. "School Choice: 'The Most Efficient' Way to Desegregate." *Connecticut Mirror*, February 2, 2010.

Frank, Thomas. *What's the Matter with Kansas?* New York: Metropolitan Books, 2004.

Friedman, Thomas L. *Hot, Flat, and Crowded: Why We Need a Green Revolution and How It Can Renew America*. New York: Farrar, Straus, and Giroux, 2008.

Galbraith, Kate. "Europe's Way of Encouraging Solar Power Arrives in the U.S." *New York Times*, March 12, 2009.

Gans, Herbert. *The Urban Villagers: Group and Class in the Life of Italian Americans*. New York: Macmillan, 1962.

Garett-Petts, William, ed. *The Small Cities Book: On the Cultural Future of Small Cities*. Vancouver: New Star Books, 2005.

Garreau, Joel. *Edge City: Life on the New Frontier*. New York: Anchor Books, 1991.

Garreau, Joel. "Today's Temples of Consumption Don't Have to Be Tomorrow's Ruins. What's in Store?" *Washington Post*, November 16, 2008, M-1.

Gearino, Dan. "Auto Industry's Pain Multiplied for Parts Suppliers." *Columbus Dispatch*, June 3, 2009.

Geddes, Patrick, "Civics: As Applied Sociology." In *Sociological Papers*. Edited by V. V. Branford. London: MacMillan, 1905.

George, Val, Colleen Matts, and Susan Schmidt. *Institutional Food Purchasing*. East Lansing, MI: Michigan State University, 2010.

Glaeser, Edward L. "Why Has Globalization Led to Bigger Cities?" *New York Times*, Economix blog, May 19, 2009.

Glanz, James. "Geothermal Drilling Safeguards Imposed." *New York Times*, January 15, 2010.

Goist, Park Dixon. *From Main Street to State Street: Town, City, and Community in America*. Port Washington, New York: Kennikat Press, 1977.

Gold, Russell, and Ann Davis. "Oil Officials See Limit Looming on Production." *Wall Street Journal*, November 19, 2007.

Goodman, Paul, and Percival Goodman. *Communitas: Ways of Livelihood and Means of Life*. New York: Columbia University Press, 1947.

Gormory, Ralph, and William Baumol. *Global Trade and Conflicting National Interests*. Cambridge, MA: MIT Press, 2001.

Gottlieb, Robert, and Anupama Joshi. *Food Justice*. Cambridge, MA: MIT Press, 2010.

Gottman, Jean. *Megalopolis: The Urbanized Northeastern Seaboard of the United States*. New York: Twentieth Century Fund, 1961.

Grant, Gerald. *Hope and Despair in the American City: Why There Are No Bad Schools in Raleigh*. Cambridge, MA: Harvard University Press, 2009.

Gratz, Roberta Brandes. *The Living City: Thinking Small in a Big Way*. New York: Simon and Schuster, 1989.

Gray, Thomas B. "The Aesthetic Condition of the Urban Freeway." Unpublished student report, University of Texas at Austin, 1999.

Griswold, Nigel G., and Patricia E. Norris. "Economic Impacts of Residential Property Abandonment and the Genesee County Land Bank in Flint, Michigan." MSU Land Policy Institute Report 2007-05. East Lansing: Michigan State University, 2007.

Gunther, Marc. "An Emotional, Economic, Social Reset." Marcgunther.com, November 6, 2008.

Hall, Douglas G., Shane J. Cherry, Kelly S. Reeves, Randy D. Lee, Gregory R. Carroll, Garold L. Sommers, Kristine L. Verdin. *Water Energy Resources of the United States with Emphasis on Low Head/Low Power Resources*. Idaho Falls: National Engineering and Environmental Laboratory, 2004.

Hall, Peter. *Cities of Tomorrow*. 3rd ed. Malden, MA: Blackwell, 2002.

Hamm, Michael. "Michigan Economic Development: Opportunities Through Local Sustainable Agriculture." Presentation at the Greentown Conference, Grand Rapids, MI, July 23, 2009.

Harper, Althea, Annie Shattuck, Eric Holt-Giménez, Alison Alkon, and Frances Lambrick. *Food Policy Councils: Lessons Learned. Report of the Institute for Food and Development Policy*. Oakland, CA: Institute for Food and Development Policy, 2009.

Haugen, Dan. "Why Isn't the U.S. Embracing Feed-In Tariffs?" *SolveClimate News*, March 24, 2009.

Hayden, Dolores. *Building Suburbia: Greenfields and Urban Growth, 1820–2000*. New York: Vintage Books, 2003.

Heilbroner, Robert L. *An Inquiry into the Human Prospect*. New York: Norton, 1974.

Heimlich, Ralph E., and William D. Anderson. *Development at the Urban Fringe and Beyond: Impacts on Agriculture and Rural Land*,—AER-803." United States Department of Agriculture Economic Research Service (June 2001).

Helper, Susan. "Renewing 'Made in the USA.'" *Washington Post*, March 4, 2008.

Helphand, Kenneth. *Defiant Gardens: Making Gardens in Wartime*. San Antonio, TX: Trinity University Press, 2006.

Henry, Fran. "Ohio's Farmers Expand into Shrimp Aquaculture." *Cleveland Plain Dealer*, August 31, 2005.

Hess, David J. *Localist Movements in a Global Economy*. Cambridge, MA: MIT Press, 2009.

Hewlett, E., and P. Melchett. "Can Organic Farming Feed the World? A Review of the Research." Paper delivered at the IFOAM Organic World Conference, Modena, Italy, June 16–20, 2008.

Hill, Jason, Erik Nelson, David Tilman, Stephen Polasky, and Douglas Tiffany. "Environmental, Economic, and Energetic Costs and Benefits of Biodiesel and Ethanol Biofuels." *Proceedings of the National Academy of Sciences of the United States of America* 103, no. 30 (2006): 11206–11210.

Hines, Eric. *Localization: A Global Manifesto*. London: Earthscan, 2000.

Hoekstra, Dave. "Rebuilding Flint: A Desolate Town Driven to Be Green." *Shore*, March 19, 2010.

Hofstadter, Richard. *The Age of Reform*. New York: Vintage, 1955.

Houston, Dan. "Local Works: Examining the Impact of Independent Business on the West Michigan Economy." Civiceconomics.com (September 2008).

Hoyt, Lorlene, and André Leroux. *Voices from Forgotten Cities: Innovative Revitalization Coalitions in America's Older Small Cities*. New York: PolicyLink, 2007.

Hui, T. Keung. "Critics Say Survey Tells Wake Board to Rethink." *Raleigh News and Observer*, February 5, 2010.

Hui, T. Keung. "Two Businessmen Invested Big in Schools Race." *Raleigh News and Observer,* February 9, 2010.

Hunt, Tam. "Why Is Community Choice Aggregation So Promising?" RenewableEnergyWorld.com, August 20, 2008.

Hurst, Blake. "The Omnivore's Delusion: Against the Agri-Intellectuals," American.com, July 30, 2009.

Hutchisson, James M. *The Rise of Sinclair Lewis, 1920–1930*. University Park: Penn State University Press, 1996.

Inslee, Jay, and Bracken Hendricks. *Apollo's Fire: Igniting America's Clean Energy Economy*. Washington, DC: Island Press, 2008.

Iverson, Joan Nassauer, ed. *Placing Nature: Culture and Landscape Ecology*. Washington, DC: Island Press, 1997.

Jackson, Kenneth T. *Crabgrass Frontier: The Suburbanization of the United States*. New York: Oxford University Press, 1985.

Jacobs, Jane. *Cities and the Wealth of Nations: Principles of Economic Life*. New York: Vintage Books, 1984.

Jacobs, Jane. *The Death and Life of Great American Cities*. New York: Vintage Books, 1962.

Jacobs, Jane. *The Economy of Cities*. New York: Vintage Books, 1969.

Kaffer, Nancy. "Some Say City Could Do More with Less." *Crain's Detroit Business*, August 23, 2009.

Kahlenberg, Richard D. "Rescuing *Brown* v. *Board of Education:* Profiles of Twelve School Districts Pursuing Socioeconomic Integration." New York: Century Foundation, 2007.

Kahn, Deborah. "Community Bids to Bypass Utilities Facing Hurdles in Calif." *New York Times*, January 7, 2010.

Kanter, James. "First Solar Claims $1-a-Watt 'Industry Milestone.'" *New York Times*, February 24, 2009.

Katz, Bruce, and Jennifer Bradley. "The Detroit Project: A Plan for Solving America's Greatest Urban Disaster." *New Republic*, December 9, 2009.

Kellogg, Alex P. "Detroit's Smaller Reality: Mayor Plans to Use Census Tally Showing Decline as Benchmark in Overhaul." *Wall Street Journal,* February 27, 2010.

Kildee, Dan, Jonathan Logan, Alan Mallach, and Joseph Schilling. *Regenerating Youngstown and Mahoning County through Vacant Property Reclamation: Reforming Systems and Right-Sizing Markets*. Washington, DC: National Vacant Properties Campaign: February 2009.

Kingsley, Thomas, and Karina Fortuny. *Urban Policy in the Carter Administration.* Washington, DC: Urban Institute, May 2010.

Klier, Thomas, and James Rubenstein. *Who Really Made Your Car?* Kalamazoo, MI: Upjohn Institute, 2008.

Kolbert, Elizabeth. *Field Notes from a Catastrophe: Man, Nature, and Climate Change*. New York: Bloomsbury, 2006.

Kotkin, Joel. *The City: A Global History*. New York: Random House, 2005.

Kotkin, Joel. *The Next Hundred Million: America in 2050*. New York: Penguin, 2010.

Kraus, Sibella. "A Call for New Ruralism." *Frameworks* 3 (Spring 2006): 26–29.

Krier, Léon. *The Architecture of Community*. Edited by Dhiru A. Thadani and Peter J. Hetzel. Washington, DC: Island Press, 2009.

Krier, Léon. "The City within the City." *Architectural Design* 54 (1984): 70–105.

Krugman, Paul. *Geography and Trade*. Cambridge, MA: MIT Press, 1992.

Kummer, Corby. "A Papaya Grows in Holyoke." *Atlantic Monthly* (April 2008): 115–118.

Kunstler, James Howard. *The Geography of Nowhere: The Rise and Decline of America's Man-Made Landscape*. New York: Simon & Schuster/Touchstone, 1993.

Kunstler, James Howard. *The Long Emergency: Surviving the Converging Catastrophes of the Twentieth Century*. New York: Atlantic Monthly Press, 2005.

Kuttner, Robert. *Everything for Sale: The Virtues and Limits of Markets*. New York: Knopf, 1997.

Lang, Robert E., and Arthur C. Nelson. "The Rise of the Megapolitans." *Planning* 73 (2007): 7–12.

Lawson, Laura. *City Bountiful: A Century of Community Gardening in America*. Berkeley: University of California Press, 2005.

Ledebur, Larry, and Jill Taylor. *Akron, Ohio: A Restoring Prosperity Case Study*. Washington, DC: Brookings Institution, 2007.

Lee, Don. "Fighting for 'Made in the USA.'" *Los Angeles Times*, May 8, 2010.

Lerch, Daniel. *Post Carbon Cities: Planning for Energy and Climate Uncertainty*. Portland, OR: Post Carbon Institute, 2007.

Leroux, André. "New England's Small Cities: A Mostly Untapped Resource." Boston: Federal Reserve Bank of Boston, 2010.

Levitt, Julia."Preservation and Sustainability: The District Approach," September 22, 2010. http://www.metropolismag.com/pov/20100922/preservation-and-sustainability-the-district-approach. Accessed March 21, 2011.

Lewis, Sinclair. *Babbitt*. New York: Harcourt Brace Jovanovich, 1922.

Lewis, Tom. *Divided Highways: Building the Interstate Highways, Transforming American Life*. New York: Viking, 1997.

Lindsay, Greg. "Demolishing Density in Detroit: Can Farming Save the Motor City?" *Fast Company,* March 5, 2010.

Lingeman, Richard. *Sinclair Lewis*. New York: Random House, 2002.

Lingeman, Richard. *Small Town America: A Narrative History, 1620–Present*. Boston: Houghton Mifflin, 1980.

Livengood, Rebecca et al. *Rethinking I-81*. Syracuse, New York: Onondaga Citizens League, 2009.

Longworth, Richard. *Caught in the Middle: America's Heartland in the Age of Globalism*. New York: Bloomsbury, 2008.

Lord, Charles, Eric Strauss, and Aaron Toffler. "Natural Cities: Urban Ecology and the Restoration of Urban Ecosystems." *Virginia Environmental Law Journal* 21 (2003): 454–551.

Luccarelli, Mark. *Lewis Mumford and the Ecological Region: The Politics of Planning*. New York: Guilford Press, 1995.

Luria, Dan D., and Joel Rogers. *Metro Futures: Economic Solutions for Cities and Their Suburbs*. Boston: Beacon Press, 1999.

Lynd, Robert S., and Helen Merrell Lynd. *Middletown: A Study in Modern American Culture*. New York: Harcourt, Brace, 1929.

Lynd, Robert S., and Helen Merrell Lynd. *Middletown in Transition: A Study in Cultural Conflicts*. New York: Harcourt, Brace, 1937.

MacDonald, Christine. "Detroit Mayor Bing Emphasizes Need to Shrink City." *Detroit News,* February 25, 2010.

MacKaye, Benton. *The New Exploration: A Philosophy of Regional Planning*. New York: Harcourt, Brace, 1928.

Mahoney, Timothy. "The Small City in American History." *Indiana Magazine of History* 99 (2003): 311–330.

Mallach, Alan, and Lavea Brachman. *Ohio Cities at a Turning Point: Finding a Way Forward*. Washington, DC: Brookings Institution, 2010.

Mallach, Alan, Subrata Basu, Stephen A. Gazillo, Jason King, Teresa Lynch, Edwin Marty, Colin Meehan, Marsha Garcia, and Erin Simmons. *Leaner, Greener Detroit*. Washington, DC: American Institute of Architects, Sustainable Design Assessment Team, 2008.

Massachusetts Department of Energy Resources. *Massachusetts Department of Energy Resources Municipal Utility Study*. Boston: Massachusetts Department of Energy Resources, 2010.

May, Henry F. *The End of American Innocence: A Study of the First Years of Our Own Time, 1912–1917*. New York: Knopf, 1959.

McCally, Karen. "The Life and Times of Midtown Plaza." *Rochester History* 69 (Spring 2007): 1–30.

McCarus, Chris. "Banking on Flint: County Treasurer Dan Kildee Collects National Attention for Land Bank." *Dome,* July 16, 2008.

McCrummen, Stephanie. "Republican School Board in N.C. Backed by Tea Party Abolishes Integration Policy." *Washington Post*, January 12, 2011.

McHarg, Ian. *Design with Nature*. Washington, DC: American Museum of Natural History, 1969.

McKenzie, Roderick D. *The Metropolitan Community*. New York: McGraw-Hill, 1933.

McKibben, Bill. *Deep Economy: The Wealth of Communities and the Durable Future*. New York: Holt, 2007.

McKibben, Bill. *Eaarth: Making a Life on a Tough New Planet*. New York: Times Books, 2010.

McKibben, Bill. *The End of Nature*. 2nd ed. New York: Anchor Books, 1999.

McMahon, Ed. *Conservation Communities: Creating Value with Nature, Open Space, and Agriculture*. Washington, DC: Urban Land Institute, 2010.

Meadows, Donella, Dennis L. Meadows, Jorgen Randers, and William W. Behrens III. *The Limits to Growth: A Report for the Club of Rome's Project on the Predicament of Mankind*. New York: Signet, 1972.

Meter, Ken, and Jon Rosales. "Finding Food in Farm Country: The Economics of Food and Farming in Eastern Minnesota." Report for Community Design Center Hiawatha's Pantry Project, University of Minnesota. Minneapolis: Crossroads Resource Center, 2001.

Michigan Food Policy Council. "Report of Recommendations Prepared for Governor Jennifer M. Granholm," October 12, 2006.

Mills, C. Wright. *The Power Elite*. New York: Oxford University Press, 2000.

Mills, C. Wright. *White Collar: The American Middle Classes*. New York: Oxford University Press, 2002.

Mills, C. Wright and Melville J. Ulmer. *Small Business and Civic Welfare: Report for the Smaller War Plants Corporation. U.S. War Production Board*. Washington, DC: U.S. Government Printing Office, 1946.

Minteer, Ben. *The Landscape of Reform: Civic Pragmatism and Environmental Thought in America*. Cambridge, MA: MIT Press, 2006.

Mishkovsky, Nadeja, Mathew Dalbey, Stephanie Bertaina, Anna Read, and Tad McGalliard. *Putting Smart Growth to Work in Rural Communities*. Washington, DC: International City/County Management Association, 2010.

Mitchell, Stacy. *Big-Box Swindle: The True Cost of Mega-Retailers and the Fight for America's Independent Businesses*. Boston: Beacon Press, 2006.

Moffat, David. "New Ruralism: Agriculture at the Metropolitan Edge." *Places* 18, no. 2 (2006): 72–75.

Mohamed, Rayman. "The Economics of Conservation Subdivisions." *Urban Affairs Review* 41 (2006): 376–399.

Mohl, Raymond A. *The Interstates and the Cities: Highways, Housing, and the Freeway Revolt*. Washington, DC: Poverty and Race Research Action Council, 2002.

Mohl, Raymond, and Arnold Richard Hirsch, eds. *Urban Policy in Twentieth-Century America*. New Brunswick, NJ: Rutgers University Press, 1993.

Moore, Adrian. "Utilities Takeover by Cities a Mistake." Reason.org, January 14, 2003, http://reason.org/news/show/utilities-takeover-by-cities-a. Accessed March 16, 2011.

Moreton, Bethany. *To Serve God and Wal-Mart: The Making of Christian Enterprise*. Cambridge, MA: Harvard University Press, 2009.

Morin, Richard, and Paul Taylor. *Suburbs Not Most Popular, But Suburbanites Most Content*. Washington, DC: Pew Research Center, 2009.

Morris, David, and Karl Hess. *Neighborhood Power: The New Localism*. Boston: Beacon Press, 1975.

Mumford, Lewis. *City Development*. New York: Harcourt, Brace, and Company, 1945.

Mumford, Lewis. *The City in History*. New York: Harcourt, 1961.

Mumford, Lewis. *The Culture of Cities*. New York: Harcourt Brace Jovanovich, 1938.

Mumford, Lewis. *The Urban Prospect*. London: Secker & Warburg, 1968.

Neuman, William. "Creativity Helps Rochester's Transit System Turn a Profit." *New York Times*, September 15, 2008.

Neumark, David, Junfu Zhang, and Stephen Ciccarella. "The Effects of Wal-Mart on Local Labor Markets." *Journal of Urban Economics* 63 (2008): 405–430.

Nolen, John. *New Ideals in the Planning of Cities, Towns, and Villages*. New York: American City Bureau, 1919.

Nolen, John. *New Towns for Old: Achievements in Civic Improvement*. In *Some American Small Towns and Neighborhoods*. Edited and introduced by Charles D. Warren. Amherst: University of Massachusetts Press, 2005.

Nolen, John. *Replanning Small Cities: Six Typical Studies*. New York: B. W. Huebsch, 1912.

Norman, Jon. "Small Cities' Fates: Population, Income and Employment Change in Smaller Metro Areas in the United States, 1970 to 2000." Paper presented at the annual meeting of the American Sociological Association, New York, August 11, 2007.

Norquist, John O. *The Wealth of Cities: Revitalizing the Centers of American Life*. New York: Basic Books, 1999.

Oehler, Kay, Stephen C. Sheppard, and Blair Benjamin. *The Economic Impact of Nuestras Raíces on the City of Holyoke: Current and Future Projections*. Holyoke, MA: Center for Creative Community Development, February 2007.

Ofori-Amoah, B., ed. *Beyond the Metropolis: Urban Geography as If Small Cities Mattered*. Lanham, MD: University Press of America, 2007.

Okrent, David. "The Tragedy of Detroit: How a Great City Fell—and How It Can Rise Again: A Special Report." *Time*, October 5, 2009, 26–35.

Oppenheimer, Mark. "Medium Town: On Living in a City Smaller Than New York." *New Haven Review* 1 (2007): 29–36.

Orfield, Gary, and Chungmei Lee. *Historic Reversals, Accelerating Resegregation, and the Need for New Integration Strategies*. Los Angeles: University of California, Los Angeles, 2007.

Owen, David. *Green Metropolis: Why Living Smaller, Living Closer, and Driving Less Are the Keys to Sustainability*. New York: Riverhead, 2009.

Paarlberg, Robert. "Attention, Whole Foods Shoppers." Foreign Policy.com, May–June 2010, http://www.foreignpolicy.com/articles/2010/04/26/attention_whole_foods_shoppers. Accessed March 16, 2011.

Painter, Gary, and Zhou Yu. *Immigrant Influx into Mid-Size Metro Areas to Influence Housing Markets*. Los Angeles: University of Southern California, Lusk Center for Real Estate, 2010.

Parsons, Kermit C., and David Schuyler. *From Garden City to Green City: The Legacy of Ebenezer Howard*. Baltimore, MD: Johns Hopkins University Press, 2002.

Patton, Wendy, with analysis by Heidi Garrett-Peltier. *The Impact of IMPACT: Creating Jobs in Ohio*. Policy Matters Ohio (February 2010).

Paulson, Amanda. "Resegregation of U.S. Schools Deepening." *Christian Science Monitor*, January 25, 2008.

Peshkov, Alex. "Holyoke Farm Program Aids Immigrants." *Springfield Republican*, February 23, 2008.

Peters, Christian J., Nelson L. Bills, Arthur J. Lembo, Jennifer L. Wilkins, and Gary W. Frick. "Mapping Potential Foodsheds in New York State: A Spatial Model for Evaluating the Capacity to Localize Food Production." *Renewable Agriculture and Food Systems* 21, no.1 (2009): 72–84.

Pew Center on Global Climate Change. *Hydrokinetic Electric Power Generation*. Washington, DC: Pew Center, 2009.

Phillips, Kevin. *Bad Money: Reckless Finance, Failed Politics, and the Global Crisis of American Capitalism*. New York: Viking Press, 2008.

Pollan, Michael. *The Omnivore's Dilemma: A Natural History of Four Meals*. New York: Penguin, 2006.

Porter, Michael. *The Competitive Advantage of Nations*. New York: Free Press, 1990.

Pozzi, Francesca, and Christopher Small. "Vegetation and Population Density in Urban Suburban Areas in the U.S.A." Paper presented at the Third International Symposium of Remote Sensing of Urban Areas, Istanbul, Turkey, June 2002.

Presad, Eswar. "The U.S.-China Economic Relationship: Shifts and Twists in the Balance of Power." Testimony before the U.S. Economic and Security Review Commission, February 25, 2010.

Priddle, Alisa. "Cracks in Auto Supply Chain Spreading." *Detroit News*, March 10, 2009.

Priesnitz, Wendy. "Counting Our Food Miles." *Natural Life Magazine* (July/August 2007): 32–33.

Procter, Mary, and Bill Matuszeski. *Gritty Cities*. Philadelphia: Temple University Press, 1978.

Radick, Lea. "How a Pioneering University Hopes to Cut Its Carbon Footprint by Half." *New York Times*, May 29, 2009.

Randolph, John, and Gilbert M. Masters. *Energy for Sustainability: Technology, Planning, Policy*. Washington, DC: Island Press, 2008.

Ravitch, Diane. *The Death and Life of the Great American School System: How Testing and Choice Are Undermining Education*. New York: Basic Books, 2010.

Reich, Charles A. *The Greening of America*. New York: Random House, 1970.

Reich, Robert. *The Work of Nations: Preparing Ourselves for 21st Century Capitalism*. New York: Knopf, 1992.

Rich, Motoko. "Factory Jobs Return, But Employers Find Skills Shortage." *New York Times,* July 1, 2010.

Robinson, Charles Mulford. *The Improvement of Towns and Cities; or, The Practical Basis of Civic Aesthetics*. G. P. Putnam's Sons, 1903.

Rosin, Nancy. *The Hands That Feed Us: 100 Years at the Rochester Public Market*. Rochester, NY: City of Rochester, 2005.

Ross, Catherine L. "Megaregions: Literature Review of the Implications for U.S. Infrastructure Investment and Transportation Planning." Report submitted by the Center for Quality Growth and Regional Development at the Georgia Institute of Technology to the U.S. Department of Transportation Federal Highway Administration, September 2008.

Rubenstein, James M., and Thomas Klier. "Restructuring of the Auto Industry." Paper presented at the Industry Studies Association Annual Conference, Chicago, May 2009.

Rugare, Steve, and Terry Schwarz, eds. *Cities Growing Smaller.* Cleveland: Urban Design Collaborative, 2008.

Rumberger, Russell, and Gregory Palardy. "Does Segregation Still Matter? The Impact of Student Composition on Academic Achievement in High School." *Teachers College Record* 107 (Spring 2005): 1999–2045.

Russo, Richard. *Empire Falls*. New York: Vintage Books, 2001.

Safford, Sean. *Why the Garden Club Couldn't Save Youngstown: The Transformation of the Rust Belt*. Cambridge, MA: Harvard University Press, 2009.

Satterthwaite, David, and David Dodman. "Are Cities Really to Blame?" *Urban World* 1, no. 2 (March 2009): 1, 12–13.

Schapiro, Julie. *Beyond the Motor City*. Blueprint America, PBS Reports on Infrastructure, 2010.

Schilling, Joe, and Jonathan Logan. "Greening the Rust Belt: A Green Infrastructure Model for Right Sizing America's Shrinking Cities." *Journal of the American Planning Association* 74 (2008): 451–466.

Schofield, Rob. "A History Lesson for Our Public Schools." NCPolicyWatch .com, February 13, 2010.

Schorer, Mark. *Sinclair Lewis: An American Life*. New York: McGraw-Hill, 1961.

Schragger, Richard. "The Anti–Chain Store Movement, Localist Ideology, and the Remnants of the Progressive Constitution, 1920–1940." *Iowa Law Review* 90 (2005): 1011–1094.

Schumacher, Ernst Friedrich. *Small Is Beautiful: Economics as if People Mattered*. New York: Harper Colophon, 1973.

Scott, Mel. *American City Planning since 1890*. Berkeley: University of California Press, 1971.

Scott, Robert E. "Unfair China Trade Costs Local Jobs." Washington, DC: Economic Policy Institute, 2010.

Scroop, Daniel. "The Anti–Chain Store Movement and the Politics of Consumption." *American Quarterly* 60 (2008): 925–950.

Sherman, Lauren. "America's Downsized Cities." *Forbes*, March 18, 2010.

Shuman, Michael. *Going Local: Creating Self-Reliant Communities in a Global Age*. New York: Routledge, 2000.

Shuman, Michael. *The Small-Mart Revolution: How Local Businesses Are Beating the Global Competition*. San Francisco: Berrett-Koehler, 2006.

Siegel, Beth, and Andy Waxman. *Third-Tier Cities: Adjusting to the New Economy*. Washington, DC: U.S. Economic Development Administration, 2001.

Singer, Audrey. *The New Geography of United States Immigration*. Washington, DC: Brookings Institution, July 2009.

Smith, Alisa, and J. B. MacKinnon. *Plenty: One Man, One Woman, and a Raucous Year of Eating Locally*. New York: Harmony/Random House, 2007.

Springer, Scott. *FoodNYC: A Blueprint for a Sustainable Food System*. New York: New York University and Just Food, February 2009.

Stratton, Emily M. *New Ruralism*. Athens: University of Georgia Land Use Clinic, 2009.

Sugrue, Thomas. *The Origins of the Urban Crisis: Race and Inequality in Postwar Detroit*. Princeton, NJ: Princeton University Press, 1996.

Sum, Andrew, Ishwar Khatiwada, and Sheila Palma. *Labor Underutilization Problems of U.S. Workers across Household Income Groups at the End of the Great Recession: A Truly Great Depression among the Nation's Low Income Workers amidst Full Employment among the Most Affluent*. Boston: Northeastern University, 2010.

Sussman, Carl, ed. *Planning the Fourth Migration: The Neglected Vision of the Regional Planning Association*. Cambridge, MA: MIT Press, 1976.

Tabb, William K. "The Centrality of Finance." *Journal of World-Systems Research* 13, no. 1 (2007): 1–11.

Talen, Emily. *New Urbanism and American Planning: The Conflict of Cultures*. New York: Routledge, 2005.

Teaford, Jon C. *Cities of the Heartland: The Rise and Fall of the Industrial Midwest*. Bloomington: Indiana University Press, 1993.

Teter, Seth. "A Shrimp Tale: How Ohio Came to Produce Fresh Seafood." *Our Ohio* 84, no. 3 (January 2006): 20–22, 24.

Thelan, David P. *Robert M. La Follette and the Insurgent Spirit*. Madison: University of Wisconsin Press, 1985.

Todorovich, Petra, ed. *America 2050: An Infrastructure Vision for the 21st Century*. New York: Regional Plan Association, 2008.

Tönnies, Ferdinand, ed. *Community and Civil Society*. Edited by Jose Harris. Cambridge: Cambridge University Press, 2001.

Tumber, Catherine. "The City's Limits." *Wilson Quarterly* 34 (Winter 2010): 108–109.

Tumber, Catherine. "Small, Green, and Good: The Role of Neglected Cities in a Sustainable Future." *Boston Review* 34, no. 2 (2009): 23–27.

UK HungerFree Campaign. "Brief on Sustainable Agriculture." October 2009.

UN Conference on Trade and Development and UN Environment Programme Capacity Building Task Force. *Organic Agriculture and Food Security in Africa*. New York: United Nations, 2008.

UN Department of Public Information. Press briefing on 2009 Revision of World Urbanization Prospects, March 25, 2010.

Unger, Nancy C. *Fighting Bob La Follette: The Righteous Reformer*. Chapel Hill: University of North Carolina Press, 2000.

Urban Land Institute. "Case Study: Prairie Crossing, Grayslake, Illinois." In *Developing Sustainable Planned Communities*. Edited by Jo Allen Gause. (Washington, DC: Urban Land Institute, 2007), 202–211.

Urofsky, Melvin I. *Louis D. Brandeis: A Life*. New York: Pantheon, 2010.

U.S. Department of Agriculture, Economic Research Service. "Farm Household Economics and Well-Being: Beginning Farmers, Demographics and Labor Allocations." http://www.ers.usda.gov/briefing/wellbeing/demographics.htm. Accessed March 17, 2011.

U.S. Department of Agriculture, Economic Research Service. "Measuring Rurality: Urban Influence Codes." Washington, DC: U.S. Government Printing Office, 2003.

U.S. Department of Labor Statistics. "Motor Vehicle and Parts Manufacturing." In *Career Guide to Industries, 2010–2011 Edition*. December 3, 2010.

U.S. Energy Information Administration. *Independent Statistics and Analysis. Electric Power Industry Overview 2007*. Washington, DC: U.S. Energy Information Administration, 2007.

U.S. Energy Information Administration. "Regional Energy Profiles: U.S. Household Electricity Report," July 14, 2005.

U.S. House Committee on Science and Technology. "Hearings on Examining Marine and Hydrokinetic Energy Technology: Finding the Path to Commercialization." Statement of Jacques Beaudry-Losique, Deputy Assistant Secretary for Renewable Energy, Office of Energy Efficiency and Renewable Energy, U.S. Department of Energy, December 3, 2009.

Vance, James. *The Merchant's World: The Geography of Wholesaling*. Englewood Cliffs, NJ: Prentice Hall, 1970.

Vey, Jennifer S. *Restoring Prosperity: The State Role in Revitalizing America's Older Industrial Cities*. Washington, DC: Brookings Institution, 2007.

Vey, Jennifer S., John C. Austin, and Jennifer Bradley. *The Next Economy: Economic Recovery and Transformation in the Great Lakes Region*. Washington, DC: Brookings Institution, 2010.

Vey, Jennifer S., and Richard M. McGahey, eds. *Retooling for Growth: Building a 21st Century Economy in America's Older Industrial Areas*. Washington, DC: Brookings Institution, 2008.

Wald, Matthew L. "War against a Wind-Rich Super Grid." *New York Times*, April 30, 2010.

Waldheim, Charles, ed. *The Landscape Urbanism Reader*. New York: Princeton Architectural Press, 2006.

Warren, Elizabeth. "The Vanishing Middle Class." In *Ending Poverty in America: How to Restore the American Dream*. Edited by John Edwards, Marion Crain, and Arne L. Kalleberg. New York: New Press, 2007.

Weaver, Clyde Mitchell, David Miller, and Ronald Deal Jr. "Multilevel Governance and Metropolitan Regionalism in the United States." *Urban Studies* 37 (May 2000): 851–876.

Whitford, David. "Can Farming Save Detroit?" *Fortune* 161, no. 1 (January 18, 2010): 78–84.

Wilbur Smith Associates. *Future Highways and Urban Growth*. New Haven, CT: Automobile Manufacturers' Association, 1961.

Williamson, Thad, David Imbroscio, and Gar Alperowitz. *Making a Place for Community: Local Democracy in a Global Era*. New York: Routledge, 2002.

Wilson, William H. *The City Beautiful Movement*. Baltimore, MD: Johns Hopkins University Press, 1989.

Wilson, William Julius. *The Truly Disadvantaged: The Inner City, Underclass, and Public Policy*. Chicago: University of Chicago Press, 1990.

Wirth, Lewis. "Urbanism as a Way of Life." *American Journal of Sociology* 44, no. 1 (1938): 1–24.

Wood, Robert. "Cities in Trouble." *Domestic Affairs* 1 (1991).

Yergin, Daniel. "Ensuring Energy Security: Old Questions, New Answers." *Foreign Affairs* 85 (2006): 69–82.

Yudelson, Jerry. "LEEDing Retail to Greener Pastures." *International Council of Shopping Centers Research Review* 16, no. 2 (2009): 54–60.

Zernike, Kate, and Megan Thee-Brenan. "Poll Finds Tea Party Backers Wealthier and More Educated." *New York Times,* April 14, 2010.

Web Sites and Blogs

1000 Friends of Wisconsin: http://www.1kfriends.org

Apollo Alliance: http://.apolloalliance.org

Benfield, Kaid, blog: http://switchboard.nrdc.org/blogs/kbenfield/

Brookings Institution Metropolitan Policy Program: www.brookings.edu/metro.aspx

Center for Community Progress: www.communityprogress.net

Center for Middletown Studies, Ball State University: http://cms.bsu.edu/Academics/CentersandInstitutes/Middletown.aspx

Center for Neighborhood Technology: www.cnt.org

Citistates Group wire service: http://citistates.com/citiwire/

Clinton Climate Initiative: www.clintonfoundation.org/explore-our-work/#/clinton-climate-initiative

Community Wealth Initiative, The Democracy Collaborative, University of Maryland: www.community-wealth.org.

Congress for the New Urbanism: www.cnu.org

COWS—Center on Wisconsin Strategy: www.cows.org

Crossroads Resource Center: www.crcworks.org

Form-Based Codes Institute: www.form-basedcodes.org

German Marshall Fund Cities in Transition Initiative: gmfus.org/cc/citiesintransition

Global Midwest Initiative, Chicago Council on Global Affairs: www.globalmidwest.org

Institute for Local Self Reliance: www.ilsr.org

Kunstler, James Howard, Web site: http://www.kunstler.com

Lincoln Institute of Land Policy: http://www.lincolninst.edu

MassInc Gateway Cities Program: www.massinc.org/INCSpot/Categories/Gateway-Cities.aspx

New Geography blog: www.newgeography.com/category/story-topics/small-cities

Post Carbon Institute: www.postcarbon.org

Rebuilding the Cities That Built America: www.rebuildingcitiesthatbuiltamerica.com/

Renn, Aaron, The Urbanophile blog: www.urbanophile.com

Rustwire: News from the Rust Belt: http://rustwire.com

Smart Growth America: www.smartgrowth.org

U.S. Department of Agriculture, Appropriate Technology Transfer for Rural Areas (ATTRA) National Sustainable Agriculture Information Service: http://attra.ncat.org

U.S. Department of Agriculture, Sustainable Agriculture Research and Education: www.sare.org

U.S. Department of Energy Database of State Incentives for Renewables and Efficiency.(DSIRE): www.dsireusa.org/

U.S. Department of Energy Information Administration, Independent Statistics and Analysis: www.eia.doe.gov

U.S. Department of Environmental Protection Office of Sustainable Communities, http://www. epa.gov/about epa/opei.html#OSC

Index

Urban and Industrial Environments

Series editor: Robert Gottlieb, Henry R. Luce Professor of Urban and Environmental Policy, Occidental College

Maureen Smith, *The U.S. Paper Industry and Sustainable Production: An Argument for Restructuring*

Keith Pezzoli, *Human Settlements and Planning for Ecological Sustainability: The Case of Mexico City*

Sarah Hammond Creighton, *Greening the Ivory Tower: Improving the Environmental Track Record of Universities, Colleges, and Other Institutions*

Jan Mazurek, *Making Microchips: Policy, Globalization, and Economic Restructuring in the Semiconductor Industry*

William A. Shutkin, *The Land That Could Be: Environmentalism and Democracy in the Twenty-First Century*

Richard Hofrichter, ed., *Reclaiming the Environmental Debate: The Politics of Health in a Toxic Culture*

Robert Gottlieb, *Environmentalism Unbound: Exploring New Pathways for Change*

Kenneth Geiser, *Materials Matter: Toward a Sustainable Materials Policy*

Thomas D. Beamish, *Silent Spill: The Organization of an Industrial Crisis*

Matthew Gandy, *Concrete and Clay: Reworking Nature in New York City*

David Naguib Pellow, *Garbage Wars: The Struggle for Environmental Justice in Chicago*

Julian Agyeman, Robert D. Bullard, and Bob Evans, eds., *Just Sustainabilities: Development in an Unequal World*

Barbara L. Allen, *Uneasy Alchemy: Citizens and Experts in Louisiana's Chemical Corridor Disputes*

Dara O'Rourke, *Community-Driven Regulation: Balancing Development and the Environment in Vietnam*

Brian K. Obach, *Labor and the Environmental Movement: The Quest for Common Ground*

Peggy F. Barlett and Geoffrey W. Chase, eds., *Sustainability on Campus: Stories and Strategies for Change*

Steve Lerner, *Diamond: A Struggle for Environmental Justice in Louisiana's Chemical Corridor*

Jason Corburn, *Street Science: Community Knowledge and Environmental Health Justice*

Peggy F. Barlett, ed., *Urban Place: Reconnecting with the Natural World*

David Naguib Pellow and Robert J. Brulle, eds., *Power, Justice, and the Environment: A Critical Appraisal of the Environmental Justice Movement*

Eran Ben-Joseph, *The Code of the City: Standards and the Hidden Language of Place Making*

Nancy J. Myers and Carolyn Raffensperger, eds., *Precautionary Tools for Reshaping Environmental Policy*

Kelly Sims Gallagher, *China Shifts Gears: Automakers, Oil, Pollution, and Development*

Kerry H. Whiteside, *Precautionary Politics: Principle and Practice in Confronting Environmental Risk*

Ronald Sandler and Phaedra C. Pezzullo, eds., *Environmental Justice and Environmentalism: The Social Justice Challenge to the Environmental Movement*

Julie Sze, *Noxious New York: The Racial Politics of Urban Health and Environmental Justice*

Robert D. Bullard, ed., *Growing Smarter: Achieving Livable Communities, Environmental Justice, and Regional Equity*

Ann Rappaport and Sarah Hammond Creighton, *Degrees That Matter: Climate Change and the University*

Michael Egan, *Barry Commoner and the Science of Survival: The Remaking of American Environmentalism*

David J. Hess, *Alternative Pathways in Science and Industry: Activism, Innovation, and the Environment in an Era of Globalization*

Peter F. Cannavò, *The Working Landscape: Founding, Preservation, and the Politics of Place*

Paul Stanton Kibel, ed., *Rivertown: Rethinking Urban Rivers*

Kevin P. Gallagher and Lyuba Zarsky, *The Enclave Economy: Foreign Investment and Sustainable Development in Mexico's Silicon Valley*

David N. Pellow, *Resisting Global Toxics: Transnational Movements for Environmental Justice*

Robert Gottlieb, *Reinventing Los Angeles: Nature and Community in the Global City*

David V. Carruthers, ed., *Environmental Justice in Latin America: Problems, Promise, and Practice*

Tom Angotti, *New York for Sale: Community Planning Confronts Global Real Estate*

Paloma Pavel, ed., *Breakthrough Communities: Sustainability and Justice in the Next American Metropolis*

Anastasia Loukaitou-Sideris and Renia Ehrenfeucht, *Sidewalks: Conflict and Negotiation over Public Space*

David J. Hess, *Localist Movements in a Global Economy: Sustainability, Justice, and Urban Development in the United States*

Julian Agyeman and Yelena Ogneva-Himmelberger, eds., *Environmental Justice and Sustainability in the Former Soviet Union*

Jason Corburn, *Toward the Healthy City: People, Places, and the Politics of Urban Planning*

JoAnn Carmin and Julian Agyeman, eds., *Environmental Inequalities Beyond Borders: Local Perspectives on Global Injustices*

Louise Mozingo, *Pastoral Capitalism: A History of Suburban Corporate Landscapes*

Gwen Ottinger and Benjamin Cohen, eds., *Technoscience and Environmental Justice: Expert Cultures in a Grassroots Movement*

Samantha MacBride, *Recycling Reconsidered: The Present Failure and Future Promise of Environmental Action in the United States*

Andrew Karvonen, *Politics of Urban Runoff: Nature, Technology, and the Sustainable City*

Daniel Schneider, *Hybrid Nature: Sewage Treatment and the Creation of the Industrial Ecosystem*

Catherine Tumber, *Small, Gritty, and Green: The Promise of America's Smaller Industrial Cities in a Low-Carbon World*